Biography

James Egan was born in 1985 and grew up in
Portarlington, Co. Laois in the Midlands of Ireland.
In 2008, James moved to England and studied in Oxford.
James married his wife in 2012 and currently lives in
Havant in Hampshire.
James had his first book, 365 Ways to Stop Sabotaging
Your Life, published in 2014.
Several of James' books have become No.1 Best Sellers
in the UK including 1000 Facts about Horror Movies,
3000 Facts About the Greatest Movies Ever, 365 Things
People Believe That Aren't True, Another 365 Things
People Believe That Aren't True, and 500 Things People
Believe That Aren't True.

Books by James Egan

Fairytale
Inherit the Earth
Words That Need to Exist in English
Hilarious Things That Kids Say
Hilarious Things That Mums Say
3000 Facts about TV Shows
3000 Facts about Animated Shows
3000 Facts about Actors
3000 Facts about Countries
Dinosaurs Had Feathers (and other Random Facts)
3000 Facts about Animals
1000 Facts about James Bond
1000 Inspiring Facts
The Pocketbook of Phobias
How to Psychologically Survive Cancer
1000 Out-of-this-World Facts about Space
3000 Facts about the Greatest Movies Ever
1000 Facts about Film Directors
1000 Facts about Superhero Movies
1000 Facts about Superheroes Vol. 1-3
1000 Facts about Supervillains Vol. 1-3
1000 Facts about Comic Books Vol. 1-3
1000 Facts about Animated Films
1000 Facts about Horror Movies
1000 Facts about American Presidents
Adorable Animal Facts
3000 Facts about Video Games
500 Things People Believe That Aren't True
1000 Things People Believe That Aren't True
3000 Astounding Quotes
1000 Facts About Comic Book Characters Vol. 1-3
100 Classic Stories in 100 Pages
Words That Need to Exist in English
The Pocketbook of Phobias
500 Facts about Godzilla
365 Ways to Stop Sabotaging Your Life

Your Teeth Can Explode

(And Other Disorder Facts)

by

James Egan

ISBN: 978-0-244-14908-6

Because of the dynamic nature of the Internet, any web addresses or links contain in this book may have changed since publication and may no longer be valid. The views expressed in this work are solely those of the author and do not necessarily reflect the views of the publisher, and the publisher hereby disclaims any responsibility for them.

Any people depicted in stock imagery provided by Thinkstock are models, and such images are being used for illustrative purposes only.
Certain stock imagery © Thinkstock.

Lulu Publishing Services rev. date: 10/01/2019

Content

Terminology

i) Condition – An illness that can be treated with medicine.

ii)   Disease – An illness that impairs normal functioning of the body.

iii)   Disorder – A functional abnormality or disturbance in a person's body.

iv)   Syndrome – A group of medical signs or symptoms that are correlated with each other.

v) Comorbidity - If a person develops more than one mental disorder, it is known as comorbidity.

vi)   Virus – An infectious agent that replicates itself inside living cells. Although it's a thousand times smaller than bacteria, viruses are dangerous as they can mutate more in one day than most animals do in millions of years. They can also fix mutations and exchange genes with other viruses.

1. Acne

AKA Acne Vulgaris

If a person's pores are clogged, they will develop pimples, usually on the face. This is known as acne. When a hair follicle is clogged, it is known as a comedo, (which is Latin for "to eat up.") If the pore is open, it will produce a blackhead. If the pore is closed, it will produce a whitehead.

Acne usually kicks in when a person hits puberty. However, some people develop acne in their 40s or 50s.

Acne is a side effect of evolution. Our ancestors had much thicker body hair than we do. Although we lost most of our body hair, we didn't lose the sebaceous glands below. The purpose of these glands is to oil the hair. Since we are missing most of our body hair, the glands are producing this oil unnecessarily, which irritates the skin, causing acne.

Several animals develop acne including rhinos, mice, and hairless dogs.

Some sources state that acne is caused by sunscreen. This only applies to certain sunscreens like Helioplex. Zinc oxide is an effective sunscreen that doesn't exasperate acne.

Sadly, there is no known way to prevent acne from developing. Washing and diet have little to no effect on it. Although many people believe spots get worse when a person is stressed, there is no conclusive evidence to confirm this. If a person regularly scratches or scrubs their spots, it can cause the skin to become permanently scarred. Using benzoyl-peroxide at night is effective at treating spots.

2. Addison's Disease

AKA Primary Adrenal Insufficiency, Hypocortisolism
This condition causes damage to the adrenal glands, which discolours the skin. Weirdly, people with Addison's usually don't look sick. In fact, many of them appear to be the peak of health since the disorder gives their skin a bronze tinge.

John F. Kennedy is often considered to be the most handsome American president. Many people say that JFK had a natural glow about him. However, JFK suffered Addison's and this "glow" was from his adrenal glands poisoning him. Kennedy's disorder was so severe, he was rushed to office nine times during his two years in office. It was so bad, Kennedy was given the last rites during one of his hospital visits.

Addison's can be treated with tablets that can mimic the effects of cortisol. People with this disorder should carry a cortisol syringe, a MedicAlert bracelet, or an information card so medical personnel can inject them as soon as possible if they are hospitalized.
Example – Jane Austen may have suffered from this illness.

3. Adermatoglyphia

Adermatoglyohia is a genetic disorder that stops a person developing fingerprints. Because this compromises a person's ability to be identified, anyone with this condition is not allowed to enter certain countries including the United States.

4. AfterImage

AKA Palinopsia, which means "seeing"
This is a condition where a person keeps seeing an

image after the stimulus has been removed. Although this can occur due to brain damage, prescription drugs, hallucinogens, seizures, or migraines, it can also happen by playing too many video games. Playing Tetris for hours on end can cause a person to see block-like images everywhere they look. This condition can last hours or even days.

5. Ageusia

Ageusia is a complete loss of taste. A partial loss of taste is known as hypogeusia.

Ageusia can be caused by nerve damage or a deficiency of vitamin B3 and zinc.

6. Agnosia

Agnosia is the inability to process sensory information. There are at least 26 types of agnosia including

i)   Phoagnosia – The inability to recognize voices.
ii) Cortical deafness – Despite the ears working perfectly, a person with this deafness can't hear because the brain doesn't send auditory signals.
iii) Astereognosis – The inability to feel anything. The villain from The World is Not Enough, Renard, suffered astereognosis.
iv) Achromatopsia – The inability to see diverse colors. An achromatopsiac usually sees things in shades of grey. Chromatopsia is the opposite of this disorder as it causes a person to see everything in overly bright colours.
v)   Mirror Agnosia – The inability to recognize how mirrors work. As a result, a person with

this disorder convinces themselves that their reflection is a stranger. Even if a person with this condition is told how mirrors work, they don't accept it.

vi) Simultagnosia – The inability to process visual input as a whole in faces, rooms, pictures, objects, people, etc.

This condition is usually associated with a brain injury, usually to the occipitotemporal border. At this moment, there is no cure.

7. Agraphia

This neurological disorder prevents a person from reading. Although there are several forms of treatment to help the patient recognize certain letters or words, there is no cure for agraphia.

Example - Howard Engel woke up one day to learn he had this disorder when he wasn't able to read the newspaper. Ironically, Engel was an author.

8. Albinism

AKA Hypopigmentation

Albinism is the congenital absence of pigmentation. This can occur in humans, animals, and plants. Human albinos have white skin, white hair, and pink eyes. Since there is no pigmentation in the eyes, albinos are more sensitive to light. The skin is white since it is not able to absorb melanin. Because of this, albinos get sunburnt twice as quickly.

1 in 20,000 people in the United States suffer albinism. Africans are nearly seven times more likely to suffer this condition than any other race.

Although albinism is a tolerable condition, the real

danger is other people. In many cultures, albinos are believed to be cursed and so, are ridiculed, beaten, or killed. Albinos are nearly always depicted as being evil in films, much to the dismay of the National Organization for Albinism and Hypopigmentation. The head of this organization, Michael McGowan, gave 68 examples of albinos being depicted as evil in films including The Da Vinci Code, Cold Mountain, End of Days, Pluto Nash, Powder, The Firm, Me, Myself, and Irene, Lethal Weapon, The Princess Bride, and The Matrix Reloaded.

Example – Actor and rapper, Krondon III, suffers albinism. He is best-known for playing the villain, Tobias Whale, in the tv series, Black Lightning.

9. Alcohol Withdrawal

AKA Delirium tremens, Saunders-Sutton Syndrome, Blue Horrors, Bottleache, Drunken Horrors, Gallon Distemper, Quart Mania, Barrel-Fever

When a person stops drinking alcohol after consuming it for a long period, there is a 50% chance they will suffer severe withdrawal. According to the National Institute of Health, an alcoholic is a person who drinks about seven pints of beer or four pints of wine daily for months at a time.

When the drinking abruptly stops, the withdrawal usually occurs three days later and will last for two or three days. During this time, the person's body should jitter, which is why it used to be called The Shakes. They may feel cold and sweaty at the same time. Their heart rate will become irregular, which can develop into a seizure in rare occurrences. The person will go through a phase of bewilderment and may suffer visual or audio hallucinations. Many people with withdrawal hallucinate spiders, bats, and elephants. These types of hallucinations are so common, the withdrawal has been nicknamed Pink Spiders, The Bats, and The Elephants. The skin will feel tingly which is why the sufferer may hallucinate that there are insects crawling around their body. A similar withdrawal occurs with certain drugs including barbiturates and benzodiazepine.

As horrific as withdrawal feels, it's easily treatable. A person going through withdrawal should be fine if they are hydrated and have plenty of rest. They should be in a well-lit room as it's more likely they will hallucinate or have nightmares if they are in darkness. In extreme cases, the individual should be cared for in an intensive care unit.

10. Alexithymia

Alexithymia is the inability to identify, show, or describe emotions. Unsurprisingly, a person with this disorder has incredible difficulty maintaining relationships with their family, friends, and partners. "Alexithymia" is Greek for "no words for emotion."

11. Alice in Wonderland Syndrome

AKA Dysmetropsia, Lilliputian Hallucinations

This is a neurological disorder that causes the sufferer to see objects far smaller or larger than they actually are. This condition can be broken into four types –

i)   Micropsia – Objects appear small
ii)   Macropsia – Objects appear large
iii)   Pelopsia – Objects appear close
iv)   Teleopsia – Objects appear far away

AIWS can develop due to a brain tumor, migraine, or hallucinogenic drugs. There is no treatment for the condition but it usually goes away over time.

12. Alopecia

Alopecia is a loss of hair on a part of the body. There are three types of alopecia.

      i) Totalis – Total loss of hair on a section of the body e.g. the head or face

      ii)   Universalis – Total loss of hair on the entire body

      iii)   Arearta – Patches of hair loss throughout the body. This is usually a symptom of a separate autoimmunity disorder. This is the most common type of alopecia. People with this arearta usually take medication to boost their immune system.

Alopecia can be caused by HIV/AIDS, chemotherapy, genetics, malnutrition, medication, or lupus.

It is possible for the hair to grow back but the growth may be irregular. A person with alopecia totalis may have one eyebrow grow back but it can fall out again several months or years later.

13. ALS (Amyotrophic Lateral Sclerosis)

AKA Motor Neurone Disease, Lou Gehrig's Disease

ALS is a rapidly progressing neurological disease that attacks the nerve cells that are in charge of controlling voluntary muscles. As the muscles shrivel up, the person will become so weak, they will have great difficulty moving, breathing, or swallowing. When the ALS develops, it usually takes a year for the infected person to be diagnosed. The sufferer normally dies three years after developing ALS. 0.002% of people develop this terrifying disorder.

Since it only attacks muscles that can be consciously moved, ALS doesn't affect the brain, meaning that it doesn't impair the sufferer's motor functions. It is unknown what causes ALS and there is no cure.

Example - Theoretical physicist, Stephen Hawking was diagnosed with ALS when he was 21 and was given two years left to live. Hawking died when he was 76, meaning that he lived with ALS longer than anyone in history.

14. Alzheimer's

Alzheimer's is a debilitating neurodegenerative disorder that causes a person to suffer short-term memory loss, mood swings, disorientation, and difficulty understanding simple concepts e.g. not recognizing a face, name, or location.

The first person to be diagnosed with Alzheimer's was Auguste Deter. When she couldn't remember her own name, Auguste repeatedly said, "I have lost myself."

When a person is about 65 years old, it's common for them to become forgetful about basic things. If this happens at an alarming rate, that person should see a doctor where they can take a cognitive test to see if they have developed Alzheimer's.

This disorder doesn't just destroy the sufferer's memory. It kills people within three to nine years. Of the 29.8 million people recorded with the illness in 2015, 1.9 million of them died that year.

Although 70% of Alzheimer's sufferers develop the disorder due to genetics, it can be caused from a head injury, depression, or hypertension. Biologically, Alzheimer's is caused by a lack of acetylcholine.

An Alzheimer's sufferer will need medication to slow down the disease. The affected person will rely on a caregiver as the disease worsens.

15. Amnesia

A person with amnesia has great difficulty recovering memories. It can be caused by brain damage, a disease, or psychological trauma.

There are several types of amnesia –

i)   Retrograde - Almost every amnesiac character in films suffers from retrograde amnesia, which causes them to forget specific events.

ii) Anterograde – The inability to generate new memories.

iii) Lacunar – The loss of memory of one specific event

iv) Korsakoff's Syndrome – The loss of memory due to long-term alcoholism

v)   Dissociative Fugue – The loss of memory due to a trauma

vi) Post-traumatic – The loss of memory due to suffering an injury, usually a head collision

In movies, a person can lose and regain their memories from a simple bump on the head. In reality, the memories may never return.

One of the strangest stories revolving around amnesia occurred to a man (who is only referred to as William) on March 14th 2005. After William had root-canal surgery at the dentist, he lost the ability to retain memories. William can't remember anything for longer than 90 minutes. After an hour-and-a-half has past, he believes it is 1:40pm on March 14th 2005 and he is in Germany ready to go to the dentist. William doesn't even know he has this condition until his wife tells him. She has placed detailed notes on his smartphone that explains to William everything he needs to know...until he forgets it 90 minutes later. Nobody knows how this happened.

Probably the most bizarre story about amnesia revolves around Naomi Jacob. Naomi suffered a lot of abuse as a child, which caused her to become a drug addict later in life. Over a period of ten years, she became bankrupt and homeless. At 32, something unprecedented happened to Naomi. She woke up to find that she had no memory of the previous 17 years of her life. In her mind, she went to bed as a 15-year-old and woke up as an adult with a ten-year-old she didn't recognize claiming to be her child.

But it gets weirder. She was 15 in 1993. Not only did Naomi wake up in a stranger's body but in a stranger's world filled with iPhones, the Internet, touch screens, and all sorts of gadgets she never knew could exist. It is believed that Naomi suffered so much trauma throughout her life, her brain reset. She went on to write 12 autobiographies and over 40 novels.

Examples – Jason Bourne in the Bourne films has retrograde amnesia. Leonard in the film, Memento, has anterograde amnesia.

16. Amok

Although "run amok" means "go crazy," many people don't realise that the phrase stems from a mental illness in Malaysia, Indonesia, Puerto Rico, and the Philippines.

"Amok" is derived from the Malay/Indonesian word "meng-amuk" which means "to make a furious and desperate charge."

Running amok tends to happen to spiritual-based tribes in Malaysia. The tribesmen believe those who develop amok have been corrupted by a tiger spirit, which causes them to lash out. Sometimes, an infected person convinces themselves they will be free of amok if they kill the person who placed the curse on them. This causes the amok sufferer to attack and kill anybody they see. Bizarrely, there are virtually no reports of women being affected by amok.

17. Anemia

Anemia is caused by a decrease of hemoglobin, which impairs the bloods ability to carry oxygen. This causes the person to feel light-headed, weak, unmotivated, and faint.

Anemia is common in young women. Although many people develop anemia when they have an iron deficiency, this isn't always the case. You can develop anemia if you lack Vitamin B12. If anemia is caused by an iron deficiency, chickpeas, spinach, and parley are recommended to boost one's iron levels.

18.   Anorexia

AKA Anorexia Nervosa, Rich Girl's Syndrome

Anorexia is an eating disorder that causes the sufferer to worry about eating and gaining weight. The word "anorexia" means "without appetite." The word was coined by Queen Victoria's personal physician, William Gull.

Many anorexics pretend they have eaten when they haven't and deny that they have a problem.

Although anorexia is associated with women, 20% of anorexics are men. From 2010-2016, hospitalization of middle-aged men from eating disorders like anorexia has increased by 70% in the US. Actors like Dennis Quaid suffered anorexia during the 1990s, showing it's not just a female disorder. Bizarrely, anorexia is more common with identical twins.

An extreme anorexic will develop body hair on the back, arms, legs, neck, or face. This hair is called lanugo. It's the same hair that covers a fetus while it is in the womb. The lanugo forms to keep the person warm since fat is normally used to generate heat.

An anorexic needs to eat a healthy amount of food but more importantly, deal with the psychological reasons that instigated the disorder. Medication tends to be unhelpful with treating anorexia.

19. Anosmia

Anosmia is the inability to perceive odor. This condition can be temporary or permanent. Some people are born with anosmia and others can develop it due to a brain injury or from a blockage in the nose. Anosmia can be treated with steroids or surgery.

Example – William Wordsworth, who is best-known for writing the poem, I Wander Lonely As A Cloud, had this condition.

20. Anosogonosia

This condition causes the sufferer to be unaware that they have an illness. A person with this disorder may be paralyzed from the waist down but they are absolutely convinced that they can walk. Anosognonosia is usually caused by damage to the parietal lobe. "Anosogonosia" is Greek for "without disease."

21. Aphantasia

A person with aphantasia has a blank imagination. If I asked an aphantasian to picture a dragon, they wouldn't be able to create an image of a reptilian winged creature with scales, claws, and teeth. Instead, they would think of the word "dragon." Although this condition was discovered in the 19th century, it wasn't taken seriously until 2015.

Although this disorder sounds frustrating, it gets worse. It's not just limited to mental images. Think of Michael Jackson's Beat It or Adele's song, Hello. An aphantasian can't think of the lyrics or the beat, even if they heard the song a million times. They also can't remember or explain a taste or smell unless they are experiencing it at that moment.

22. Arthritis

This disorder causes joint pain, swelling, and stiffness. 24.9% of women and 18.1% of men suffer arthritis, making it one of the most common disorders in the world. In fact, it is the leading cause of disability in the United States. Arthritis is more common with African Americans but uncommon with Hispanics.

Although there are approximately one hundred types of arthritis, the most common types are –
i) Osteoarthritis breaks down joint cartilage. Osteoarthritis usually affects the hips, hands, and knees but can also affect the ankles and feet.
ii) Rheumatoid arthritis causes the joints to break down when the affected person rests. Bizarrely, a 2010 study from Brazil showed that bee sting venom was surprisingly effective at combating rheumatoid arthritis.

Although it is seen as an "old disorder," people in their 20s and 30s can develop arthritis. There are cases where children have developed the disorder.

Basic medication like ibuprofen and paracetamol are effective at dealing with the pain but they don't slow down the disorder. Weight loss, exercise, and joint replacement are the most effective treatments for arthritis. Depending on the type of pain, a heat pack or ice pack can be useful.

Foods with Vitamin E and Omega 3 fatty acids have anti-inflammatory properties which can combat against arthritis. Although it is treatable, there is no known cure for arthritis.

23.   Arctic Hysteria
AKA Pibloktoq
Arctic hysteria causes a person to become deranged when they have experienced intense cold for a long period. This disorder usually happens to Inuits. Because an Inuit's diet contains a lot of fat and very little Vitamin A, they can become aggressive, disorientated, and unbearably hot. This causes the sufferer to tear off all their clothes, even in sub-zero

temperatures, and run around frantically. Inuits believe the erratic behaviour is caused by an evil spirit.

24.   Asperger Syndrome

Asperger's is a developmental disorder that causes a person to mentally develop at a slow rate, causing them to have trouble with communication and social interaction. A person with Asperger's usually speaks with another person in a one-way communication. They may want to talk about their hobbies but will not react to anything the other person says. This gives the false impression that Asperger sufferers are emotionless. However, people with Asperger's are far more emotional than a person without the disorder.

Although many people mix up Asperger's with autism, they are not the same thing. Although every autistic person has Asperger's, not every person with Asperger's has autism.

There are three key differences –

i)   Asperger's sufferers are usually of normal or high intelligence while an autistic person is of lower intelligence.

ii)   Although a person with Asperger's may have impaired speech, an autistic has little to no verbal capabilities.

iii) Autism is noticeable by the time a child is two. Asperger's is usually not noticeable until a child is in school.

Although Asperger's seems to be on the rise, this is only because doctors are much better at diagnosing the disorder. It's incredibly likely that Isaac Newton, Thomas Jefferson, and Albert Einstein had Asperger's but the disorder wasn't recognized at the time.

Nobody knows what causes Asperger's. Although it can be treated with social gatherings, therapy, education, and medication, there is no cure.

25.   Asthma

Asthma is an inflammation of the airway of the lungs. It is the most common childhood disease in the world. Asthmatics regularly suffer from a chesty cough, tight chest, and breathlessness. Most asthma sufferers have a trigger such as dust, smoke, allergens, pollen, mold, tobacco, temperature change, overexertion, illness, or stress.

Asthma is diagnosed and measured with a peak flow meter. Blowing into this device will show if the lungs are inflamed.

Most asthmatics use a medical device called an inhaler to help with their breathing. An inhaler is filled with corticosteroids, which reduces lung inflammation.

Because this inhaler is so accessible in the modern world, society forgets how terrifying asthma is. If an asthma sufferer didn't take this inhaler regularly, it's possible that he or she will die. Many asthma sufferers in Third World countries have no access to inhalers and so, countless die every single day. In fact, about 397,100 people die from asthma attacks every year.

Asthma's severity varies from person to person. One person can have mild asthma which clears up when they reach adulthood. Another person can have asthma for a lifetime and it affects them on a daily basis.

26.   Attention Deficit Hyperactivity Disorder

Until recently, many ADHD sufferers were accused of being lazy or stupid. However, it's more complicated than that. There are three types of ADHD –

i)   Type I are predominately hyperactive. This is what most people think of when they hear of ADHD.

ii)   Type II are predominantly inattentive. This is more common with females. It's sometimes referred to as an invisible disability because it's very difficult to detect. Just because a Type II sufferer isn't jumping around screaming doesn't mean they don't have incredible difficulty being attentive.

iii)   Type III – Combination of Type I and II

A person with ADHD is believed to be easy to distract. Sometimes, it's the exact opposite. A person with ADHD can be so focused on something, they may not notice something incredibly obvious like a fire. This state of mind is called Hyperfocus. This is where the idea comes from that people with ADHD are inattentive or prone to day-dreaming.

ADHD can be treated with behavioral therapies, medication, or something as basic as diet.

27. Autism

Autism is a bio-neurological developmental abnormality which causes the individual to suffer physical and mental impairments. This means that autism is caused by problems in the person's neurons, not their DNA. This is important as it means that autism is not hereditary and so, cannot be passed on. It also proves that you can't "get" autism from an outside agent like a vaccine since it cannot rewrite a person's neurons.

When autism was classified in 1943, it was perceived as a disorder that made a person relate to objects more than people.

Autistics tend to have habits or "rituals" which they regularly perform such as swinging their left arm every time they enter a room, tapping a specific key on a piano, or rubbing their head every time they see an orange car. These rituals can make the autistic person's life difficult. If their ritual is only eating beige-colored food, it can be troublesome to convince them to eat fruit or vegetables.

The disorder is usually diagnosed when a child is three or four years old.

Although the number of autistics seems to have increased drastically over the years, this is because society is far better at identifying the disorder.

Although males seem to suffer autism almost five times more often than girls, neurologists have recently concluded that this estimate may be inaccurate. Autism in girls seems less common because their habits are more socially acceptable than males.

At birth, there is no difference in brain anatomy between those with autism and those without. However, when an autistic person reaches their first year, their amygdala grows 13% larger. The amygdala is responsible for emotionally reacting. Although autistics have great difficulty expressing emotion, they are more emotional than someone without autism. Occupational therapist, Lindsey Biel, who cares for people with mental difficulties has stated that autistics are very expressive when they write down how they feel.

Recently, studies have found a new way to detect autism in children. When a person smells an unpleasant odor, it's natural for them to try not to inhale. However, autistic people don't do this. An autistic person will breathe in a foul stench just as much as a pleasant odor. The study's co-author, Liron Rozenkrantz, said The Sniff Test can detect if a child has autism in ten minutes. Although the test is 100% accurate, it has received positive results.

Interestingly, autistic people can tickle themselves. *Example* - Although the film, Rain Man, popularized the idea that autistic people are mathematical geniuses, this only applies to very few. People with this ability are known as autistic savants. However, most savants are non-verbal.

Autistic savants like Daniel Tammet can recite pi to 22,514 decimal places or learn a language in a week but cannot dress himself or make a cup of coffee.

28. Auto-Brewery Syndrome
AKA Gut Fermentation Syndrome

Imagine being pulled over by a police officer for suspected drunk-driving. Imagine you blow into the breathalyzer that verifies you are way over the limit.

Now imagine that you have never drunk alcohol in your entire life. ABS is a disorder that causes ethanol to develop in the digestive system, which causes a person to test positive for a breathalyzer even if they haven't drunk alcohol. Unsurprisingly, many drunk-drivers have pretended to have this disorder when they have been pulled over by the police. It's very difficult to convince anyone you have this disorder since it emulates drunkenness. ABS causes fatigue, slurred speech, confusion, aggression, dizziness, belching, anxiety, IBS, and a hangover.

ABS can be treated with a change of diet involving low carbohydrates and high protein.

29. Avoidant Personality Disorder

AVD causes a person to be petrified of criticism. They obsess about their limitations, failures, and feelings of inferiority. A person with AVD is so scared of being humiliated or disliked, he or she becomes socially inept. AVD is usually a side effect of anxiety or agoraphobia. Although AVD can develop due to emotional neglect or an abusive childhood, the disorder can be genetic.

Group therapy, drug therapy, and social skills training is effective at reintroducing an AVD sufferer into society.

30. Bearded Lady Syndrome
AKA Hypertrichosis

This disorder causes the person to grow excess body hair. It's different from hirsutism because the hair grows in normal places, but at an incredible rate. Women with this disorder usually develop a beard.

For centuries, people with hypertrichosis appeared in freak shows as a human/ape hybrid or a werewolf. Although the hair can be shaved off, it grows back at astounding speed.

There are cases of women growing beards due to conditions besides hypertrichosis, including polycystic ovaries.

Example – Lettie Lutz in the film, The Greatest Showman, was based on a bearded lady called Anna Jones.

31.   Benjamin Button Syndrome
AKA Werner Syndrome, Progeria

This is a genetic disorder that causes children to suffer premature aging. The word "progeria" is Greek for "before old age."

In the 1996 film, Jack, Robin Williams' character suffers progeria, so he looks 40 years old even though he's ten. In reality, the disorder doesn't work this way.

Usually, progeria-sufferers have normal intelligence but have withered bodies due to their cells aging too quickly. They suffer hair loss, stunted growth, fragile bones, and pinched facial features.

Sadly, sufferers of this illness have a life expectancy of only 13 years. This disease is caused by a mutation during embryonic development. Although scientists know how it occurs, they don't know how to stop it or treat it.

32.   Berserk Llama Syndrome

This condition causes a llama to believe a human is another llama. Usually, the llama will become aggressive and start attacking the human. This disorder is terrifying and hilarious at the same time.

33.   Bibliomania

Although bibliophilia is a love of literature, bibliomania is a dangerous obsession with books. Not only do bibliomaniacs buy, steal, and hoard books, they usually don't have any interest in reading them. A bibliomaniac might buy several copies of a rare book just so nobody can else have it. This disorder has recently become common with comic book collectors. Example - Sir Thomas Phillipps was an English antiquary that owned in excess of 160,000 books.

34. Bipolar

AKA Clinical Depression, Manic Depression

Bipolar is a chemical imbalance in the brain. There are many chemicals that help the mind function. Having too much or too little of any one those chemicals can make one prone to paranoia, anxiety or aggression.

When someone is considered to have "bipolar," it usually refers to Bipolar II. Bipolar II is also known as manic depression, although this is an outdated term. It is no longer called manic depression because people see "clinical depression" as "normal depression" and "manic depression" as "super depression." Manic depression isn't necessarily worse than clinical depression; it's just a different type of illness.

Bipolar I sufferers have one heightened mood. This means the person may suffer from intense dread, anger, or sadness for seemingly no reason. The sensation can last for hours, days, or weeks.

Bipolar II sufferers have two intense mood swings. This makes Bipolar II sufferers come across like they have two personalities. As a result, Bipolar II sufferers are inaccurately perceived as having a split-personality disorder. Bipolar II sufferers have a moment of mania where they feel hyperactive, energetic, and overly happy.

But their mood will inevitably plummet, and they will feel overcome with misery, fear, sadness, or frustration. The majority of bipolar sufferers have mood swings every three to five days.

Bipolar II sufferers tend to have obsessions over the most innocuous things like fixing a clock or knowing the capital of a country. This is known as Goal Directed Activity and can last for hours or days. This is depicted very accurately in the film, Silver Linings Playbook, when a bipolar sufferer wants to discuss a book he is reading in the middle of the night while his parents are trying to sleep.

The most important thing to remember is that bipolar isn't normally triggered by external stimulus. Happy things don't necessarily make the bipolar sufferer happy and sad things don't necessarily make them sad. They can have pangs of depression for apparently no reason.

A person with this illness might be sitting down, doing nothing of significance and suddenly feel overwhelmed by happiness. When they are having an episode, it's very difficult for any external stimulus to affect their mood, for better or for worse.

They can win the lottery today but feel upset because the bipolar has set in their mind.

Bipolar affects 2.6% of the adult population. It is most common in the US, affecting 17% of the population. The illness is usually genetic and has a 71% chance of effecting an individual if one of his or her parents had bipolar.

However, the illness can develop from a series of tragic events or a trauma. It is quite treatable with the correct medication.

Unmedicated bipolar sufferers are inclined to self-harm (especially people with Bipolar I.)

35.   Bird Flu
AKA Avian Flu
Humans can catch this illness if exposed to infected bird feces or if the secretions from an infected bird are absorbed into the mouth or eye. There is no record of the virus being spread from human to human. It has almost the same symptoms as the flu but to a higher degree. 84% of birds that have been diagnosed with bird flu were farm animals.

There are two main types of bird flu –

i)   The H5N1 strain is lethal to birds and can pass to humans and other animals easily. Although it was discovered in humans in 1997, the virus didn't become well-known until it spread throughout Asia in 2003. In 2005, H5N1 made its way to Europe. Of the 630 confirmed human cases of H571 from 2003-2013, the WHO confirmed that 375 died.

ii)   The H7N9 strain is less likely to be passed onto humans but is far more lethal. Of the 775 confirmed human cases of H7N9 as of 2016, the WHO confirmed that 316 of them died. These mainly occurred in China as well as Hong Kong, Taiwan, and Malaysia. However, both strains of bird flu will only kill toddlers, the elderly, and people with weak-immune systems. There is no record of a reasonably healthy person dying from bird flu.

Bird flu can be diagnosed with a chest x-ray or using an auscultation test to detect irregular breathing.

Vaccines against bird flu have been available since 2013.

36.   Blind Love
AKA Identity Annihilation
Blind love is when a person gives up their entire life to be with someone they are infatuated with. Although this behavior is considered to be romantic, it is a recognized disorder.
Example - In The Little Mermaid, Ariel gives up her entire life, her family, her home, and her voice to be with a man she barely knows.

37.   Blindness
Despite what most of society believes, only 10-15% of blind people see absolutely nothing. Many legally blind people can decipher colors, shapes, and varying degrees of light. When the first nuclear weapon was tested in Trinity, a blind woman called Georgia Green, asked her brother, "What's that light?" even though she was 50 miles away from the blast. One of the greatest ballerinas of all time, Alicia Alonso, lost her sight when she was 19. Despite her handicap, Alonso danced into her 70s by using the spotlights on stage so she knew where to move.

A person is classified as legally blind if they have 10% eyesight or less.

Most people lose their sight gradually from age. The three most common forms of blindness are –
i)   Cataracts cause clouding in the lenses of the eyes, which develops blurry vision, glare, double vision, and faded colors. Specific

glasses are needed when a person develops cataracts.

ii)   Glaucoma damages the optic nerves, causing a person's visual ability to shrink. Glaucoma is 15 times more common in Africans than Caucasians. Prescription eyedrops and surgery can treat glaucoma.

iii)   Macular Degeneration creates blank spots in one's vision. Horror writer, Stephen King, suffers macular degeneration. Regular eye exams can allow an optometrist to gauge a specific combination of minerals and vitamins to slow down the macular degeneration.

Blind people suffer four times more nightmares than sighted people. Some blind people suffer Non-24-Sleep-Wake Disorder which causes them to have little to no sleep structure. They can sleep three hours one night but they might sleep 12 hours the following afternoon.

There's a riddle from 1688 that states, "If a man born blind can feel the differences between shapes such as spheres and cubes, could he, if given the ability, distinguish those objects by sight alone?" This riddle was solved in 2003 when five blind people held objects of different shapes and then were asked which was which after they had their sight restored. None of them could do it just by looking at the objects.

38.   Blue Skin Syndrome
AKA Argyria, Argyrosis
If a person consumes too much silver, it will impair their kidney function and night vision. However, this

condition is best-known for turning the person's skin blue. Once the skin changes color, it cannot be reversed.

Example - Paul Karason took colloidal silver, believing it would benefit his health. After several months, he developed blue skin.

39.   Boanthropy
This is a psychological disorder in which a person believes he is a bovine (buffalo, yak, antelope, cow) or is turning into one.

Example – In the Book of Daniel, Nebuchadnezzar II "was driven from men, and did eat grass as oxen."

40.   Bobble-Head Doll Syndrome
This disorder causes a person to bob their head up-and-down or side-to-side. Luckily, it stops when the patient is asleep. It tends to occur in three-year-olds. This disorder is caused by a cyst in the brain and can be treated easily with surgery.

41.   Borderline Intellectual Functioning
BIF is when a person has an intellectual disability but they are not classified as having Asperger's or autism.

Example – The titular character in Forrest Gump

42.   Brain Aneurysm
AKA Cerebral Aneurysm
Many people believe a brain aneurysm occurs when blood pumps through a weak artery in the brain, causing it to swell until it ruptures. However, a brain aneurysm is simply a weak wall in the brain. You can have an aneurysm without the artery rupturing. In

fact, 9% of Americans have had an aneurysm but most are oblivious to it! You can go your entire life not knowing you have an unruptured aneurysm. Usually, this blockage is benign and a doctor would advise not treating it as it could risk irreparable damage.

Luckily, there are warning signs before the artery ruptures. Prior to the aneurysm bursting, a patient should feel a blinding headache right behind the eyes. They will also feel a tingly sensation in their face and nape. Just before an aneurysm, the person will usually become hypersensitive to light. People who have suffered from this rupture stated they felt like they had been hit by lightning. If you suffer any of these symptoms, go to a hospital immediately. This rupture is so severe, up to 50% of patients die within the next 30 days. Up to 20% die before they arrive in hospital.

43.   Breast Hypertrophy
AKA Macromastia, Gigantomastia.
This disorder overdevelops breast tissue, causing the breasts to expand dramatically.
Example - Annie Hawkins-Turner developed this disorder when she nine years old. She has the world record for the largest natural breasts, weighing a total of 85lbs. Each breast weighs more than an average four-year-old child.

44.   Brittle Bone Disease
AKA Osteogenesis Imperfecta, Lobstein Syndrome, Fragilitas Ossium, Vrolik Disease
This genetic disorder prevents collagen from building up in the body. This causes the bones from ossifying correctly, making them easy to break. Although BBD

sufferers have undeveloped muscles, weak bones, and loose joints, the most noticeable physical traits are that the whites of the eyes are tinted blue.

There are eight types of BBD; Type I being the least severe and Type VIII being the most severe.

Although there is no cure, BBD sufferers are recommended not to smoke since it will break down collagen faster, making the bones even more fragile.
Example – Elijah Price in the films, Unbreakable and Glass.

45. Broca's Aphasia
This disorder makes a person have a very limited vocabulary despite the fact that their intelligence is completely unimpaired.
Example – Hodor from the HBO series, Game of Thrones.

46. Broken Heart Syndrome
AKA Takotsubo Cardiomyopathy
Although many people think it's just a fairytale, dying from a broken heart is possible. A sudden stressful experience (a break-up, death, rejection, etc.) can weaken muscles in the heart, causing it to give out.

47. Bubonic Plague
AKA Black Death, The Pestilence, The Great Mortality
The bubonic plague is caused by the Yersinia pestis bacterium, which is normally carried by fleas. A person will develop flu-like symptoms within one to seven days of being exposed to the bacterium. At first, the infected person suffers fever, headaches, and vomiting. Shortly after, swollen lymph nodes will

appear where the bacterium entered the skin. The lymph nodes can bulge so much, they can burst open. Eventually, the body is covered with black boils that ooze pus and blood.

Although the plague was recorded 2,600 years ago in China, it is infamous for sweeping across Europe between 1348-1353, killing up to 60% of the entire continent. At the time, many Europeans believed that the plague came from Asia and it had wiped out the entire population of India. In reality, this plague originated from marmot squirrels from Mongolia. When the bacterium passed onto rat fleas, the plague had mutated so it had become highly contagious and lethal to humans. The first known people to have died from this strain of plague were two Russians; Kutluk and his wife, Magnu-Kelka.

Contrary to popular belief, the bubonic plague is still around today. Thanks to good sanitation, it only kills a few hundred per year.

48.   Bulimia

Bulimia is an eating disorder that compels a person to binge-eat and then purge the food from their body, either through vomiting or laxatives.   The word "bulimia" is Greek for "ox hunger."

Although bulimia is associated with anorexia, bulimics are usually average or above-average weight according to the EDF. Although anorexics avoid food, the average bulimic binge-eats 11 times per week.

If a bulimic regularly vomits by pushing their fingers down their throat, it's likely the stomach acid will cause their knuckles to harden. The markings on the knuckles is known as the Russel's Sign.

Regularly vomiting can cause cracks in the corners of the mouth. These cracks are called cheilosis. Habitual vomiting over several years can cause the enamel of the teeth to erode.

Some bulimics don't purge with laxatives or vomiting. They may purge through excessive exercise or starvation.

Although the average age for a bulimic is 20, children as young as six can develop the disorder.

The amount of pressure that bulimia puts on the body can cause the patient to develop esophageal rupturing, gastrointestinal problems, and kidney failure. If a bulimic is pregnant, it heightens the chances of miscarriage, stillbirth, or birth defects.

Unlike anorexia, medication is very effective with bulimia. Cognitive therapy and dialectical behavior therapy are also advised.

Example – Paula Abdul was bulimic for 17 years. Sharon Osbourne has been bulimic for over 35 years. Russel Brand became bulimic when he was only 11.

49.   Burning Mouth Syndrome

BMS is a condition where a sufferer feels like they've burned their mouth. This can occur if the person is eating something cold or not eating at all. Although 50% of BMS sufferers are postmenopausal women, no one is sure what causes it. BMS is treated with antidepressants and anxiolytics.

50.   Cancer

Cancer is an abnormal growth of cells. As the cells grow larger, they interfere with the functionality of the organs. Metastasis is when the cancer starts to spread to other organs. This is the final stage of cancer and occurs shortly before the person's death.

If the cancer doesn't spread, it becomes a benign tumor, which is harmless.

There is a 33% chance you will develop cancer (although it's slightly more likely for men.)

Although there are over a hundred types of cancer, 22% of cancer deaths are related to tobacco. In fact, there are 69 chemicals in a cigarette that can cause cancer.

The average cancer sufferer survives for five years after their diagnosis, but it depends on what type of cancer they have. The most common cancer for women is breast cancer but it also has the highest rate of recovery. The most lethal cancer is pancreatitis, which can kill a person several weeks or months after their diagnosis. There is almost no example of a human being surviving pancreatitis.

You can develop cancer from toxic agents, radiation, auto-immune disease, or genetics.

Healthy lifestyle and diet are the best ways to avoid getting cancer. However, 5-10% of cancers are genetic so you may be susceptible to it even if you take every precaution imaginable. You might develop lung cancer despite never smoking.

Red wine is proven to kill cancer cells. Also, trained dogs can sniff out prostate cancer with 98% accuracy.

Chemotherapy and radiotherapy are the most effective treatments for cancer. Depending on the type of cancer, surgery, and laser therapy may also be effective.

Although cancer is terrifying, society is getting

better at treating it every day. In 1991, 215 people died of cancer for every 100,000 people. In 2010, 172 people died from cancer for every 100,000 people.

51. Capgras Delusion

Capgras Delusion is when a person believes a living thing has been replaced with an imposter. This condition tends to happen with paranoid schizophrenics but can develop in a patient due to a brain injury, dementia, or a neurodegenerative disease. According to The Psychiatric, Psychogenic and Somatopsychic Disorders Handbook, it's much more common in women.

Example – Christine Collins is the most famous person to be diagnosed with Capgras Delusion. Ironically, she didn't have the disorder! Although Collins' son was kidnapped, the authorities returned him to his mother soon after. However, Collins was certain the boy was not her son. Although doctors tried to convince her that she was delusional, it was later revealed that the boy wasn't actually her child. Sadly, Christine never reunited with her son.

This story was depicted in the film, Changeling, with Angelina Jolie playing Christine.

52. Cataplexy

You know when a possum plays dead in the presence of a predator? It turns out that the possum isn't "playing." When a possum is terrified, it becomes paralyzed with fear. This condition is known as cataplexy. Cataplexy, which is Greek for "stroke down," can occur in people if they are terrified or traumatized. Cataplexy can only be treated with medication, not behavioral therapy.

53. Catatonia

Catatonia is where a person has impaired motor mobility. It is a syndrome connected to autism, psychosis, bipolar, and schizophrenia. There are many different versions of catatonia including –

i) Stupor – No physical activity or reacting to one's environment. This is the syndrome most people are familiar with.

ii) Mutism – Little to no verbal communication

iii) Negativism – Oppose all communication

iv) Grimacing – Unconsciously pulling faces or making jerky movements

v) Echolalia – Mimicking sounds or speech

vi) Echopraxia – Mimicking another person's movements

vii) Catalepsy – Rigidly maintain a held position e.g. holding out a hand

viii) Waxy Flexibility – Maintain a held position but another person can move it easily

ix) Posturing – Sudden movements which will then be maintained for hours or days

x) Mannerisms – Exaggerated movements for normal actions

xi) Stereotypy – Repetitive, OCD-like habits

xii) Agitation – Spontaneous rage not based on external stimuli

Treatment usually involves a large dose of benzodiazepine or electroconvulsive therapy.

Example – The 1990 film, Awakenings, which stars Robert De Niro and Robin Williams, revolves around doctors trying to wake catatonic patients. It is based on the work of Dr. Oliver Sacks.

54.   Cerebrovascular Atherosclerosis
This causes calcium deposits to build in the cerebral arteries in the brain until it becomes solid. Basically, this condition turns a person's brain to stone!
Example – It is believed that Soviet Russian leader, Vladimir Lenin, died from this illness.

55.   Charles Bonnet Syndrome
AKA Visual Release Hallucinations
This is a visual hallucination that blind people suffer. The disorder is quite common with people who just lost their sight and can last for up to 18 months.
    Although philosophical writer, Charles Bonnet, first descried the disorder in 1760, it wasn't introduced into psychiatry until 1982.

56.   Chickenpox
AKA Varicella
Chickenpox is a highly contagious infectious disease that causes the person to develop itchy blisters throughout their body. When a person is exposed to pox, it incubates for two to three weeks before symptoms become noticeable. An infected person is most contagious a day or two before the pox is visible.
    Although it's irritating, chickenpox is rarely dangerous if the person develops it as a child. Once a person has chickenpox, it's incredibly rare for them to develop the disease again. However, if a person gets chickenpox as an adult, he or she can become infertile or develop shingles. Because of this, parents usually encourage their children to get chickenpox as soon as

possible so they will have a high immunity to the disease for the rest of their life.

If a child has chickenpox, do not give them aspirin as he or she may develop Reye's Syndrome, which damages the liver and brain.

If a pregnant woman develops chickenpox, it can cause the fetus to suffer birth defects.

Most children receive the chickenpox vaccine when they are a year old. They get their second dose when they are five years old.

57. Chimerism

This is when a twin dies in the womb and the other twin absorbs its DNA. If a person is born this way, they are known as a chimera. One chimera called Taylor Muhl has had autoimmune issues her entire life because her body keeps trying to fight off her twin's "foreign" cells. What's more bizarre is Muhl's complexion is different on both sides of her body. Half her body is pink while the other half is tanned. She looks like she is two people fused into one...which she kind of is.

58. Cholera

Cholera is an infection in the small intestine instigated by the vibrio cholerae bacterium. Cholera causes the person to have such intense diarrhea, they can die from dehydration hours after being exposed to the bacterium. However, it can take up to five days before a person feels the effect of cholera after being exposed to the bacterium.

The bacterium is found in contaminated food and unclean water, which is why cholera is common in poverty-stricken regions.

A physician in London called John Snow determined that a contaminated water pump was responsible for spreading cholera. A pub named after John Snow stands beside this pump today.

Although cholera is not as common nowadays thanks to advances in sanitation, it still infects up to five million people per year. In 2015, cholera was recorded to have killed 28,800 people.

59. Chronic Obstructive Pulmonary Disease
AKA Emphysema, Chronic Bronchitis

This lung disease causes the person to suffer long-term breathing problems. Although COPD isn't universally known, it is one of the leading causes of death in the US. When COPD first develops, the symptoms are mild so it's easy to dismiss it. If an elderly person develops a shortness in breath, they may assume it's just their age catching up on them rather than a fatal disease. The disease can fester for years and people tend to only go to a doctor when it's too late.

It tends to occur to people who are over 40 years old. Although COPD is usually caused by smoking, genetics and air pollution also must be factored in. If COPD is caught early, it can be treated. Sadly, there is no cure.

60. Claustrophobia

A claustrophobic is terrified of enclosed spaces or places that make them feel trapped. The word "claustrophobia" is Greek for "shut in fear place." It is often triggered by elevators, crowded areas, MRI machines, and windowless rooms. Being locked in a small car can also cause a claustrophobic to panic. When claustrophobia sets, it causes the person to suffer anxiety or a panic attack. They may believe that they can't breathe, move, or the room is gradually getting smaller.

Claustrophobics have a reduced amygdala, which impairs their emotional responses in stressful situations.

It is one of the most common phobias as it affects about 6% of modern society. Strangely, a Brit is twice as likely to suffer claustrophobia than an American.

61. Clinical Lycanthropy

This delusion causes the sufferer to believe they can transform into a werewolf or some sort of demonic beast. Clinical lycanthropy is usually connected with schizophrenia.

62. Cold

The common cold is a viral infectious disease caused by the rhinovirus. It causes the sufferer to develop a cough, a fever, a runny nose, and a sore throat. The average adult develops two or three colds per year, making it the most common infectious disease.

Many people believe you can catch a cold if you kiss someone infected with the virus. This is impossible since the virus is in the nasal mucus, not the mouth.

One archaic treatment many people use to treat a cold is to sweat it out. Putting a blanket on top of the head over a bowl of hot water feels nice but is utterly ineffective against a cold. Garlic, honey, lemon, and blueberries are more effective at fighting a cold than cough medicine or antibiotics.

63.   Cold Sores
AKA Herpes iabialis
A cold sore is an infection on the lip caused by the herpes simplex virus. 75% of people carry this virus, so they will suffer a cold sore at some point in their life. 20% of people routinely develop cold sores. Cold sores are contagious so it's vital not to share personal items like cloths, cups, and toothbrushes with a person who has one.

Although it starts as a burning blister, it can develop into a sore throat and a fever if is not treated quickly. Blueberries and coconut oil are effective against cold sores.

64.   Cold Urticaria
A person with cold urticaria can't tolerate cold. Now, there's plenty of people who can't stand it when the temperature is freezing. But a person with this disorder can suffer hives, an intense rash, systemic shock, or die if they feel the slightest chill.

There are many different medications for cold urticaria but a person with this condition needs to be tested to figure out which one is the most effective.

65.   Color Blindness
A person with color blindness has an impairment with deciphering one or multiple colors. Most people with this disorder have red-green color blindness. Facebook CEO, Mark Zuckerberg, is red-green color blind, which is why he made Facebook have a blue

color template.

Color blindness is divided into four categories – slightly, moderate, strong, and absolute. A person with absolute color blindness can only see 20 hues. By comparison, a person with regular vision can see about a hundred hues.

About 8% of men suffer this condition while only 0.5% of women suffer it.

Although color blindness may sound like a mild inconvenience, it can jeopardize a person's ability to get certain jobs. You need to have normal vision to become a police officer, a fireman, or a pilot. In some countries, you are not allowed to have a driver's license if you have color blindness.

Although there are different tests to see if a person is color blind, the anomaloscope is considered to be the most accurate.

Although there are corrective glasses to help an individual see more colors, there is no cure.

Example – Bill Clinton, Christopher Nolan, Eddie Redmayne, Keanu Reeves, Mark Twain, Meat Loaf, Prince William, and Vincent Van Gogh.

66. Coma

A coma is a state of unconsciousness where a person does not react to stimuli. A person who has remained unconscious for over six hours is considered comatose. A coma can be caused by many things including a central nervous system disease, a collision, stroke, hypothermia, or neurosurgery. 40% of comas are caused by drug poisoning.

Contrary to popular belief, a comatose person can speak, interact, recognize their surroundings, and walk around. But since the patient is still in a "coma," they can get back into bed, fall into a deep unconsciousness, and have no memory of any of the events they experienced.

Doctors use the Glasgow Coma Scale to gauge coma progress from a score of three (which is the rating for deep unconsciousness) to 15 (full consciousness) based on eye, verbal, and motor responses.

Elaine Esposito was in a coma for 37 years and 111 days, which is the world record. However, Esposito never woke from her coma. The longest time to awake from a coma was 19 years.

It's important to speak to a comatose patient as they can hear things around them. However, the patient will forget nearly everything they heard when they wake up.

How to awaken a comatose patient depends on the cause of the coma. If the coma was caused by cardiac arrest, lowering the patient's body temperature is effective at awakening them. However, there is no universal fix for a coma.

67. Complex Regional Pain Syndrome
AKA Reflex Sympathetic Dystrophy

CRPS causes a person to be covered in a burning body-covering rash after receiving any injury, no matter how minor. If you bump your toe, your pain receptors will send a message to your brain, notifying you of the injury. People with CRPS will have their pain receptors sending the message to the brain for hours, days, or weeks. This can make a stubbed toe or a small bruise unbearable.

68. Compulsive Punning
AKA Foerster's Syndrome

That's right. Regularly making puns can be classified as a disorder. This disorder was first recognised in 1929 when the neurosurgeon, Otfrid Foerster, removed a tumour from a patient's brain, which caused the patient to burst out laughing and ramble a series of puns. The syndrome was documented by a journalist called Arthur Koestler. Koestler found it bizarre that this syndrome was identified "from a man tied face-down to the operating table with his skull open."

69. Concussion
AKA Mild Traumatic Brain Injury

A concussion is a head injury that temporarily impairs the brain's functionality. Symptoms usually include dizziness, fatigue, blurred vision, headaches, impaired memory or concentration, nausea, and mood swings. These symptoms can last up to four weeks. Women are more likely to suffer a concussion and it takes twice as long for them to recover. It's common for a person to have a concussion and not realise it.

Concussions are often caused by vehicle collisions and sport injuries. There is less than a 10% chance that a concussion will cause the affected person to pass out.

Suffering several concussions can develop into a horrific disorder called Chronic Traumatic Encephalopathy. It is also known as Dementia Pugilistica. CTE causes dementia, aggression, depression, and suicidal behavior. CTE is most common with boxers, American footballers, and rugby players. Henry VIII's paranoia, mood swings, memory loss, and tantrums sound like a textbook example of CTE.

Example – The 2015 film, Concussion, tells the true story of pathologist, Dr. Bennet Omalu, who learned that head collisions causes brain damage after observing the autopsy of a football player.

70.   Conduct Disorder

CD compels a person to destroy property and show aggression to people and animals. CD sufferers have a tendency to compulsively lie and steal. It is often seen as a precursor to antisocial personality disorder.

Example - Stewie Griffin from the animated series, Family Guy

71.   Conjunctivitis

AKA Pink Eye, Madras Eye

Conjunctivitis is an inflammation on the white of the eye and the surface of the eyelid, which causes the area to be itchy. It is caused by an allergy or a bacterial or viral infection.

Although it is instinctive to rub your eyes when it stings, this will exasperate the problem. If viral conjunctivitis becomes serious, it can be treated with acyclovir.

72.   Coprographia

This disorder compels a person to draw obscene drawings. "Coprographia" literally means "Crap Drawing."

73. Copropraxia

This is a condition that compels a person to make inappropriate gestures. "Copropraxia" means "Crap Action."

74. Crack Baby Syndrome
AKA Prenatal Cocaine Exposure

During the 1980s, crack cocaine was a serious problem in America. Even worse, there were many reports that pregnant women smoking crack were passing on the addiction to their unborn children. The New York Times suggested that there would be so many "crack babies" born, they wouldn't be able to function in society and it would collapse the American education system.

In case you didn't notice, this never happened. Do you know why? Because crack babies don't exist! The idea that babies can be born addicted to crack and remain addicted for the rest of their lives was based on a single study of 23 infants, which was published in 1985 in The New England Journal of Medicine. However, the subjects were *only* studied as babies, so the idea that they grew up to be crack addicts is unfounded.

75. Crazy Sickness
AKA Grisi Siknis

This illness occurs among the Miskito people in Nicaragua. Sufferers will fall into a coma-like state for hours or days, only to awaken in a rage, seemingly attacking an invisible threat. There was a report of 60 Miskitos developing Crazy Sickness and attacking hundreds of people in their community. Although the tribe believe the illness is a curse, it was concluded in the 1950s that CS was caused by hallucinogens that were found in the tribe's water supply.

76. Crohn's Disease
AKA Regional Enteritis
This disease inflames the gastrointestinal tract, which can cause abdominal pain, fever, arthritis, fatigue, anemia, skin rash, weight loss, and diarrhea.

It tends to occur in people between the ages of 20-30. The disease is mainly caused by genetics. However, smokers are twice as likely to suffer the illness.

Although there is no cure, there are multiple treatments for Crohn's. Dietary adjustments and vital. Smoking is out of the question. Crohn's sufferers should avoid larger meals and food with omega-6 fatty acid and polyunsaturated fat. Bizarrely, chemotherapy has proven effective against Crohn's.

77. Cyclical Vomiting Syndrome
CVS causes a person to vomit uncontrollably with no warning. This can last a lifetime and nobody knows what causes it.
Example - Charles Darwin suffered this disorder for most of his adult life.

78. The Dancing Disease
In 1518, there was a dancing epidemic in Strasbourg. 400 sufferers danced for a month until most of them died from heart attacks. Nobody knows what caused it.

79. Deafness

A deaf person has little to no hearing. A total lack of hearing is called anacusis.

The most common reason that people lose their hearing is simply from listening to loud noises. It can be a loud noise over a period of years e.g. concerts, city noise, occupational noise, etc. or it could be a single incident e.g. an explosion.

People with hearing loss can benefit from hearing aids and cochlear implants.

80. Decompression Sickness

AKA The Bends, Caisson Disease, Diver's Disease

Decompression Sickness occurs when dissolved gases come out of bubbles inside the human body. There are two types of decompression

i)   Type I temporarily irritates the muscles, skin, and lymphatics. It mainly irritates joints in the elbows and shoulders. Most people have experienced this while swimming. If the person is experiencing this, they should lie on their side and rest until it passes.

ii)   Type II affects the ears, lungs, and brain, which can cause nausea, vomiting, fatigue, headaches, numbness, tingling, chest pains, coughing, and tinnitus. In extreme cases, the person can develop paralysis, impaired vision, fall into a coma, or die. Many divers experience this sickness 6-48 hours after diving.

You can experience decompression sickness if you are on a plane that isn't pressurized correctly. In September 2018, Jet Airways forgot to depressurize their plane cabin, causing 160 passengers to experience decompression for 45 minutes. This caused 30 passengers to develop splitting headaches and nose bleeds.

If a person experiences severe decompression sickness, they will need hyperbaric oxygen therapy and IV fluids.

81.   Delusional Infestation
AKA Delusional Parasitosis, Ekbom's Syndrome
This illness causes the sufferer to believe their body is infested with insects. The person may interpret any mark on their body as a bite from these "insects," meaning that any scratch or bruise may exasperate their paranoia. The sufferer may unintentionally hurt themselves in an attempt to get rid of the fictitious insects.

82. Dementia
This horrific disease degenerates the person's brain, which affects their ability to think and remember. They also suffer mood swings and an impaired ability to be understood or to understand others.

Although dementia was a recognized disease in Ancient Greece, it wasn't a categorized illness until 1901 when Alois Alzheimer learned that there was a separate type of dementia called Alzheimer's. Although Alzheimer's is the most common type of dementia, the disease takes on other forms -

   i) Dementia with Lewy Bodies affects the person's behavior, cognition, and movement. It usually doesn't affect the person's memory but they usually suffer horrific hallucinations, especially after waking. Comedy legend, Robin

Williams, was diagnosed with this disorder shortly before his death.

ii) Vascular Dementia is caused when the brain isn't supplied with enough blood. This usually leads to mini-strokes, which get worse over time.

iii) Frontotemporal Dementia causes the person to lose social awareness and suffer poor impulse control. However, people with FTD don't suffer problems with memory, balance, spatial awareness, or perception.

83. Depersonalization Disorder

This mental disorder causes a person to feel like they are disconnected from one's self. A person with depersonalization disorder may feel like they are an observer watching himself or herself. DPD tends to occur when a person is trying to deal with a sudden trauma or a terminal disease.

Although this sounds like a rare disorder, it occurs to 50% of Americans at least once in their life. Also, women are twice as likely to suffer DPD. Although it was first recognized during the 19th century, DPD became more recognized when war veterans said they felt like they were out of their own body.

Although medication and therapy are advised, DPD usually corrects itself over time.

84. Dermatillomania

AKA Excoriation Disorder

This is a mental disorder that compels the sufferer to pick at their own skin.

85. Diabetes

Diabetes is a number of metabolic disorders that increases a person's blood levels for an extended period of time. Insulin is created by the beta cells in the pancreas. If these cells malfunction, the person will not be able to manufacture insulin and thus, will develop diabetes. This causes the person to suffer from thirst and hunger regularly. They will also need to urinate more often. If the blood levels are not regulated, the person can become incredibly ill or die. A diabetic can measure their blood levels with a portable meter.

60-70% of diabetics develop blockages in the cardiovascular system, which can lead to deadened nerves. If this nerve damage is severe, the diabetic may have one of their limbs amputated.

Although it was first recorded in Ancient Egypt in 1500 BC, diabetes wasn't named until about 120 AD by Greek physician, Aretaeus. At the time, diabetes was diagnosed by tasting the patient's urine. Since a diabetic's urine tastes sweet, Aretaeus called it "diabetes mellitus," which means "honey flowing through."

However, Aretaeus believed diabetes was caused by a snakebite. It wasn't known until 1889 that diabetes originated from the pancreas. Oskar Minkowski discovered this when examining the pancreas of a diabetic dog. In 1921, Charles Best and Frederick Banting learned that diabetes was caused by an imbalance of insulin.

There are three types of diabetes –

i)   Type I – If beta cells don't form properly in the pancreas, the body won't be able to manufacture insulin, causing the person to develop diabetes. Type I is normally genetic. Caucasians are the most likely to develop Type I Diabetes. Type I diabetics regulate themselves with insulin injections after every meal.

ii) Type II – This is when insulin forms properly but the cells don't respond to it appropriately. Type II is usually caused by obesity, lack of exercise, and genetics. African-Americans are over twice as likely to develop Type II than Caucasians, Asians, or Hispanics. The death rate is 27% higher for diabetic African-Americans. Although Type II diabetics can use injections to regulate their blood sugar, there are many oral medicines available as well.

iii) Gestational diabetes – GD occurs with pregnant women. GD can disappear shortly after the woman has given birth.

Examples of Type I – Halle Berry, Theresa May
Examples of Type II – Tom Hanks, Larry King, Randy Jackson, HG Wells,
Examples of GD – Salma Hayek

86.   Dissociative Identity Disorder
AKA Multiple Personality Disorder
Dissociative Identity Disorder causes a person to develop a separate personality from their own, usually to combat a stressful situation. If a DID sufferer called Alice gets too stressed at work, she might switch to her other personality, Anna, who is more diligent and assertive. This other personality is called the Alter. The Alter can have completely different posture and speak in a different voice or accent. A violent Alter is called a

Persecutor. Persecutors rarely lash out at others. Usually, they direct it upon the host.

This disorder was originally called Split Personality but this was changed since society mixed up DID with schizophrenia (which means "split mind.")

DID tends to occur in people who have suffered a trauma when they were nine years old or younger. Women are more likely to develop DID as girls are more prone to suffering abuse as a child.

Although therapy is helpful for treating DID, there is no cure or medication that combats the condition.

Example - Billy Milligan was a criminal with 24 different personalities, each with their own intonation pattern and accent. Milligan was the inspiration for Kevin in the film, Split. In that film, Kevin's personalities speak to each other coherently, which is inaccurate according to clinical psychologists.

The most extreme version of DID is Kim Noble, who had a hundred personalities.

87. Down Syndrome

There are three types of Down Syndrome –

i) Trisomy 21 – 95% of people with Down Syndrome have this type. Trisomy 21 causes Chromosome 21 to split into three instead of two, causing the person to have 47 chromosomes instead of 46. This extra genetic material leads to irregular developments in the body.

ii) Translocation – This is when Chromosome 21 breaks off during cell division and merges with another chromosome. This affects 4% of people with Down Syndrome.

iii) Mosaicism – This is when Chromosome 21 fails to separate correctly during cell division after fertilization. This affects 1% of people with Down Syndrome.

People with Down Syndrome have cognitive delays, meaning that it takes them longer to pick up on things from a very early age. A baby with Down Syndrome will take longer to roll, crawl, stand, and walk.

The most striking physical attribute of a person with Down Syndrome is an upward slant in the eyes. However, there are many other characteristics including short stature, flattened nose, abnormal teeth, short neck, disproportionally small hands, low muscle tone, small chin, abnormal outer ears, and a single crease in the center of the palm. They usually have an overly large tongue and a narrow mouth. Because they have less room in their mouth, they tend to have slurred speech. They usually have narrower nasal passages, which causes them to have thicker mucus.

Down Syndrome is the most common genetic condition in the world. One in 691 babies in the US are born with Down Syndrome.

People with Down Syndrome are more prone to developing epilepsy, thyroid diseases, heart defects, and depression. They have a weak immune system so can become very sick from minor illnesses.

Life expectancy has skyrocketed for Down Syndrome sufferers in the last few decades. In 1983, the life expectancy of someone with Down syndrome was only 25. In 2018, the life expectancy was 60.

A child's skeleton was discovered in France in 1989 that belonged to a person with Down Syndrome.

The skeleton dates back to the fifth or sixth century, making it the oldest Down Syndrome remains found.

Although the exact cause of Down Syndrome is unknown, a 2018 genome sequencing study in China on 141,000 women showed that it is linked to a person's ability to have twins. It is also linked to a person's likelihood to develop herpesvirus 6.

88.   Dr. Strangelove Syndrome
AKA Alien Hand Syndrome
People with this condition can wake up in the night to learn that they are being attacked by their own hand. This hand can punch, scratch, and try to gouge out the person's eyes.

DSS usually affects people who have their brain hemispheres surgically separated. It can also affect people who have developed epilepsy, psychosis, a stroke, or Alzheimer's. Sadly, there is no cure.
Example - In the film, Dr. Strangelove (or How I Learned to Stop Worrying and Love the Bomb,) the titular character's hand keeps moving against his will.

89.   Drapetomania
Not only is this disorder not real, it's utterly immoral. In 1851, physician, Samuel A. Cartwright, published a study to justify why slavery was acceptable. Since the Bible states that slaves should obey their masters, Cartwright believed that an unwilling slave must be mentally ill. Cartwright called this illness drapetomania. However, Cartwright states that white slave-owners suffered from this disorder as well if they treated their slaves as equals.

90. Dry Drowning

This disorder is exactly what it sounds like – drowning on land. If a person inhales a mouthful of water, it can cause them to drown if the liquid reaches the lungs. The scary thing is this can happen hours or days after a person consumes water.

91. Dwarfism

A dwarf is a person who suffers a hormone irregularity, causing them to be short in stature. If an adult is shorter than 4ft 10, he or she is classified as a dwarf.

Although there are over 200 forms of dwarfism, 80% of dwarfs have achondroplasia, which stops the cartilage from ossifying into bone. Game of Thrones actor, Peter Dinklage, has this type of dwarfism.

Dwarfism is genetic so it is possible to pass it on to one's children.

According to the Guinness World Records, a Nepalese man called Chandra Dangi is the world's shortest man, standing 21.5 inches.

92. Dyscalculia

Dyscalculia is a learning disorder which gives the sufferer great difficulty understanding mathematics. No matter how hard a dsycalculiac tries, he or she may never be able to understand rudimentary math such as algebra, arithmetic, or long division.

Dyscalculia is seen as the mathematical version of dyslexia. Like dyslexia, dyscalculia has nothing to do with a person's intelligence. A dyscalculiac may not have any difficulty understanding other subjects like physics or chemistry.

11% of dyscalculiacs suffer ADHD. It is also common with people with Turner Syndrome or spina bifida.

93. Dyschronometria

Dyschronometria causes a person to have an impaired concept of time. It is very common with dementia sufferers.

Since dyschronometria is caused by damage to the cerebellar ataxia, dopamine and serotonin receptors have been proven to be effective at treating the condition.

94. Dysgeusia
AKA Parageusia

Dysgeusia distorts a person's sense of taste. This can make ice-cream taste like motor oil. It can also make non-foods taste amazing so paint might taste like a cheeseburger. Dysgeusia is a common symptom to chemotherapy as many chemo patients say that food and water taste metallic. However, dysgeusia can also be caused by a severe zinc deficiency.

95. Dyslexia
AKA Reading Disorder, Alexia

Dyslexia is a learning disability that causes the affected person to have great difficulty reading. "Dyslexia" is Greek for "inadequate language."

It is the most common learning disability, affecting 10% of the population. It is one of the few disabilities that affects males and females equally.

Although dyslexia was believed to be a sign of slow-learning until the 90s, the disability has no effect on a person's intelligence.

Unsurprisingly, many dyslexics struggle to memorize phone numbers, addresses, and foreign languages.

This condition doesn't affect the eyes of the dyslexic but the phonologic module of the brain. This impairment causes the dyslexic to see letters irregularly. There are dozens of types of dyslexia including –

i) Phonological Dyslexia – This is the most well-known and common form of dyslexia. A person with this disability had difficulty reading words.

ii) Rapid Naming Deficit – Difficulty listing things.

iii) Surface Dyslexia – Difficulty saying words that don't sound the way they are spelt e.g. debt, freight, sword, etc.

iv) Double Deficit Dyslexia –Difficulty forming specific sounds.

v) Dysgraphia – Difficulty writing or typing.

vi) Neglect Dyslexia – Skips specific sounds, letters, or words.

vii) Attentional Dyslexia – Combines sounds or words.

viii) Visual Effects – Letters appear to change e.g. a "b" changes into a "d." Sometimes, dyslexics can mix up entire words that don't seem to resemble each other e.g. "they can read "meat," but their brain may process it as "drill."

Dyslexics are naturally very creative. They excel in conceptualization, spatial ability, and thinking outside the box, which is why up to 35% of business entrepreneurs have dyslexia.

Despite what the Church of Scientology states, there is no cure for dyslexia. There are exercises to treat it such as the Orton-Gillingham Approach but there is no known way to rectify it.

Example – Albert Einstein, Alexander Graham Bell, Winston Churchill, Tom Cruise, Hans Christian Anderson, Mozart, Thomas Edison, Benjamin Franklin.

96. Dysmorphia

Dysmorphia causes a person to believe that parts of their body or appearance is dramatically different to how it actually appears. A dysmorphic can be skinny but believe their belly or their entire body is disproportionally large. Although a dysmorphic may have a big belly, he or she may believe it is much bigger than it is. According to a study committed in 2010 by Dialogues in Clinical Neuroscience, 2.4% of people are dysmorphic. Although dysmorphia is usually associated with anorexia, some people with the condition believe they are too skinny when they are visibly overweight. Others may believe they are incredibly muscular when they are clearly out of shape. If a person believes they have a full head of hair when they are going bald, it's seen as a mild case of dysmorphia.

Although dysmorphia is usually caused by abuse, neglect, denial, or peer pressure, it can be genetic.

Dysmorphics tend to have repetitive behaviors, similar to an OCD sufferer.

Cognitive behavior therapy is effective at treating dysmorphia.

97. Dysosmia

Dysosmia is the inability to smell correctly. A dysosmiac may smell garbage when they sniff a flower. They may smell chocolate when they sniff rotten meat. The scent may always match the object being sniffed or it might change every time.

There is a similar disorder called troposmia, (also known as parosmia or cacosmia) where the affected person can only smell foul odors.

There is no cure without removing the person's ability to smell.

98. Dyspraxia

AKA Developmental Coordination Disorder

This chronological neurological disorder prevents messages to the brain being transmitted correctly, which impairs the person's co-ordination. Dyspraxia is common with people suffering cerebral palsy, autism, Parkinson's, and multiple sclerosis.

Dyspraxics usually have difficulty with time-keeping, spatial ability, and sensory overload. Because dyspraxics have this disorder for life, they tend to adapt to their shortcomings, which is why they have tendency to be creative and thinking outside the box.

Since they have impaired co-ordination, dyspraxics have a tendency to trip, bump into things, or drop stuff. Although people find dyspraxics funny because they seem clumsy, this can make the sufferer feel stupid or depressed. Dyspraxia is a very tiring disorder. When a normal person gets up the morning, they can have a shower, get dressed, have breakfast, do their hair, brush their teeth, and leave the house without concentrating. Although your brain isn't fully awake, you can perform these actions because your brain is on auto-pilot. Dyspraxics need to be consciously aware while performing the most basic tasks, which can be exhausting. Although 10% have mild dyspraxia, 0.2% of people have dyspraxia that is so debilitating, they can't function in society.

Although dyspraxia is often treated with cognitive behavioral therapy or occupational therapy, there is no cure.

99. Earworm
AKA Sticky Music, Brainworm, Stuck Song Syndrome, Involuntarily Musical Imagery

Earworm is when you get a tune stuck in your head. A 60-year-old woman in the US had a song stuck in her head that was so unbearable, she had to go to the hospital. Although everyone gets a song stuck in their head at some point, this woman had to put up with it for three weeks nonstop. The woman couldn't sleep because she could hear the song as if it was being played on a recorder. When it did eventually stop, it was replaced by another song that she couldn't get out of her head. In the end, she had to take anti-seizure pills to stop the earworm from happening. This is the only case where someone had to receive medical attention for earworm.

100. Electromagnetism Hypersensitivity
Although this isn't a real illness, there are some people who claim to feel pain when exposed to

electromagnetic waves from any technology. Sweden is the only country where Electromagnetism Hypersensitivity is classified as a genuine illness.
Example – Chuck in the Netflix series, Better Call Saul, suffers dizziness and pain if he is exposed to any technology.

101. Elephantiasis

This disorder causes a person's body parts to become grossly oversized. Although there are many different versions of elephantiasis, its normally instigated by an obstruction in the lymphatic vessels, which causes a certain body part to fill with subcutaneous tissue and fluid.

Medication and surgery are required to alleviate the growth.

Contrary to popular belief, Joseph "The Elephant Man" Merrick suffered from Proteus Syndrome, not elephantiasis.

102. Ellis-van Creveld Syndrome
AKA Chondroectodermal Dysplasia

EVCS causes the sufferer to develop dwarfism and an extra thumb on each hand. Bizarrely, this condition is common with the Amish.

103. Environmental Tilt

This visual perception disorder causes a person to see their surrounds at a tilted angle. The environmental tilt can be perceived as a 90-degree or 180-degree tilt, meaning that the affected person might see everything upside down. Although these hallucinations usually last a few seconds, environmental tilts can be lethal if a person experiences one while they are driving. Sometimes, the tilt can last up to an hour.

These hallucinations are usually caused by brain injuries, migraines, strokes, or vertigo. The best treatment for a person with this disorder is to simply close their eyes until their vision resets.

104. Epilepsy

AKA Falling Sickness, The Sacred Disease

Epilepsy is a group of neurological disorders characterized by the sudden development of seizures. The most noticeable symptom from a seizure is vigorous shaking. However, a seizure can manifest many strange behaviors. While having a fit, the epileptic may speak gibberish, repeat a word or phrase, scream, or undress.

The seizure's physical traits can be so subtle, that medical experts may not notice that an epileptic is having a fit. While an epileptic is having a fit, they might just stare into space while passerbys are oblivious they are having a seizure.

Although there are different kinds of epilepsy, 60% of epileptics suffer Single Tonic-Clonic seizures. These seizures cause the sufferer to spasm for 5-10 minutes. After this time has passed, the epileptic should be fine.

If the seizure lasts longer than ten minutes, the epileptic should go to the hospital for treatment. Epilepsy can be treated with medication, diet, neurostimulation, or surgery. If left untreated, epilepsy can be fatal. Approximately 125,000 people die per year due to suffering injuries during a fit. Although the disorder can be treated, there is no cure.

The seizure has a beginning, middle, and end. The beginning is known as the Aura. This is when an epileptic can feel a seizure is imminent. They feel lightheaded or believe they can smell, hear, or taste something.

The middle part is the seizure itself, which is known as the Grand Mal. If you see an epileptic in this phase, move any objects in the person's vicinity to stop them from hitting something or hurting themselves.

The end of the seizure is known as the Postictal Phase. During this time, the brain is recovering so the sufferer may develop disorientation and memory loss for minutes or hours. An epileptic will have no memory of the Grand Mal phase.

The first book to detail epilepsy was Hippocrates' work, On the Sacred Disease, in 400 BC. Although many people at the time believed neurological disorders were a sign of possession, Hippocrates saw epilepsy as a treatable illness.

It can be caused by a brain injury, stroke, a tumor, birth defects, or an infection. However, the cause for 70% of epileptic seizures is unknown.

Many people believe an epileptic will have a fit when exposed to flickering light. Only 5% of epileptics are photosensitive. Epileptics can have fits for numerous reasons such as stress, fatigue, or allergies.

Until the 19th century, society believed that epilepsy was contagious. Toward the beginning of the 20th century, several US states forbade epileptics to marry or have children. Until recently, it was believed an epileptic can swallow their own tongue during a fit.

The original method for dealing with an epileptic seizure was to place a small book in the person's mouth to stop them from biting their tongue. This is incredibly dangerous since the victim can bite down on the book so hard, they can break their jaw or knock their teeth out.

Some people develop seizures even if they don't have epilepsy. Alcohol or drug withdrawal, a fever, low blood sugar, or a concussion can cause a person to develop a seizure.

105. Exploding Head Syndrome
AKA Episodic Cranial Sensory Shock
EPS causes a person to hear deafening noises that don't exist. This can affect the person when they are asleep, causing them to wake up to a sound that isn't real.

106. Exploding Teeth Syndrome
During the 1800s, there were several cases of people's teeth exploding. From the late 19th century onward, there has never been a case of this happening since. Researchers aren't sure why this happened but there is a theory - Teeth fillings during that time were composed of lead, tin, and silver, which could potentially create a low-voltage electrochemical cell. If these fillings came into contact with hydrogen (which could form from a cavity,) it could create an electrical charge, causing a mini-explosion.

However, this theory hasn't been tested... probably because test subjects don't have exploding teeth in their skulls.

107. Face Blindness
AKA Prosopagnosia

A prosopagnosiac has great difficulty recognizing faces. A sufferer of this condition can look at a picture of their family, having no idea who they are.

Some sufferers cannot recognize their own face. These people have Self-Recognition Prosopagnosia.

There are others who cannot recognize basic objects like a chair or an apple. This is known as Object-Discrimination Prosopagnosia.

Some people believe this condition is connected to Asperger's but this is untrue. Some people on the autistic spectrum have difficulty recognizing emotions, not faces.

Although sufferers can be born with this condition, it normally develops due to a trauma, Parkinson's, Alzheimer's, or carbon monoxide poisoning.

Example - Brad Pitt suffers this disorder and relies on recognizing people based on their scent.

108. Fainting
AKA Syncope

A person will pass out if their blood pressure is too low because the heart won't be able to pump a normal supply of oxygen to the brain. The clinical term for low blood pressure is hypotension.

Fainting can be caused by abnormal heart rhythms, dehydration, anxiety, depression, or hypoglycemia (low blood sugar.)

One of the most well-known types of fainting is vasovagal syncope, which is when a person faints upon seeing blood. Interestingly, fainting on the sight of blood has an evolutionary advantage. When a person passes out, it lowers the blood pressure. If a person had a severely wound, they could die from losing too much blood. If a person blacks out, the blood will travel around the body much slower, making them lose far less.

Regularly fainting can be avoided by a change of diet or not overexerting yourself. In extreme cases, the person may need a pacemaker.

109.   Fanboy Frenzy
AKA Lisztomania, Liszt Fever

Fanboy frenzy occurs when a person becomes hysterical when in the presence of a celebrity. The most well-known variation of fanboy frenzy is Beatlemania, when fans of The Beatles cried hysterically and passed out while in the presence of the famous British band.

Fanboy frenzy can be very dangerous. Austin St. John, who is best known for playing the Red Ranger in the tv series, Mighty Morphin Power Rangers, was stopped at a mall by fans. This crowd was so excited to see St. John, they tore his clothes off. St. John had so many encounters like this, he moved house twice and installed sensor lights on his home.

110.   Fantasy Prone Personality
AKA Maladaptive Daydreaming

This is a mental disorder that forces a person to daydream to escape from bad situations. It is an unconscious defence mechanism and the daydream is so intense, the person may seem like they are in a coma.

Example – The main character in the film, Brazil.

111. Fear Immunity,
AKA Urbach-Wiethe Disease, Lipoid Proteinosis, Hyalinosis Cutis et Mucosae
This genetic condition creates lesions on the skin, making it harder to heal. However, the main symptom of UWD is the hardening of the amygdala of the brain, which causes the person to become immune to fear. Less than 500 people have been recorded with UWD.

112.   Fetus in Fetu
AKA Absorbed Twin, Parasitic Twin
This is a developmental abnormality that causes a fetus to grow inside a body. The fetus usually forms in the carrier's abdomen. Since the fetus doesn't have a heartbeat, doctors can completely miss it in a patient during an inspection. This fetus is usually harmless and may not affect its carrier for decades. Although the fetus can develop arms, legs, vertebrae, hair, and teeth, it's not really alive. It's just a mass of cells rather than a living being. This disorder is twice as likely to occur to males. There have been less than a hundred cases of fetus in fetu ever recorded.

When a fetus in fetu is discovered, it should be surgically removed as soon as possible.
Example – In 1999, Sanju Bhagat was rushed to hospital due to a growth in his abdomen, which was believed to be a tumor. His belly was so big, Bhagat looked nine-months-pregnant.

While the surgeon cut him open, he felt a hand inside Bhaghat's body. Bhagat had a fetus in fetu growing inside him for decades. The twin was probably the most developed fetus in fetu in history since it had form half a body including genitalia, a jaw, limbs, hands, feet, a large amount of hair, and fully-formed fingernails. After the twin was removed, Bhagat made a complete recovery.

113. Fever
AKA Pyrexia, Febrile Response

When a person's body temperature is above the average 37.1 degrees Celsius (98.6 degrees Fahrenheit,) he or she will develop a fever. A fevered person usually has a temperature of 37.5-38.3C (99.5-100.9F.) As the fever sets in, the person will shiver with cold. Over time, they will feel flushed and begin to profusely sweat.

Developing a fever is good for children as it kills bacteria and viruses. On top of that, it strengthens the child's immune system. As a result, it's better not to give a feverish child any medication. The best thing to do when you have a fever is to wait it out, get plenty of rest, and stay hydrated. Having a bath with tepid water alleviates the effects of a fever.

There's an old wives' tale that you should starve a fever. This is counter-productive. You need food to gain strength when you are ill. Starving your body is going to make the fever last longer.

A fever can develop from an infectious disease, a skin inflammation, or tissue damage.

114. Fire Disease
AKA Suppressed Anger Syndrome, Hwa-byung

This disorder is specific in Korea. The illness causes intense rage in people, especially middle-aged women. Now you might think, "Isn't that just an anger management problem rather than a mental illness? Surely this happens everywhere and not just Korea, right?" Well, Fire Disease has very specific symptoms. The sufferer doesn't just get angry. Many Koreans who developed this condition said they felt a boiling sensation in their chest, blurred vision, paranoia, migraines, dizziness, fatigue, and depression. It usually happens to women who suppress their emotions but then snap during an intense incident such as learning their husband is cheating on them or they are being treated unfairly at work.

115. Fish Odor Syndrome
AKA Trimethylaminuria
If you suffer from this... I am so sorry. Fish Odor Syndrome causes a person to smell like rotting fish. Forever. You know what's worse? There is no cure. Bathing doesn't make a difference. There's no medication to fix it. Nobody knows what causes this illness and it can trigger at any point in a person's life.

116. Florence Syndrome
AKA Stendhal Syndrome, Hyperkulturemia
This psychosomatic disorder causes a person to faint if they see something awe-inspiring. It tends to affect people who are looking at art, historic sites, or natural phenomenon.

117. Flu
AKA Influenza, Gripper
The flu is an infectious disease caused by the influenza virus. The flu causes a person to develop a runny nose, cough, sore throat, fatigue, a headache, and muscle

pains. The person usually develops symptoms two days after being exposed to the flu and normally lasts a week. The flu is most contagious within the first four days of the symptoms developing.

Contrary to popular belief, you can't catch the flu from a flu-shot.

Although there have been four flu pandemics since 1900, the most famous one occurred in 1918. To avoid a panic, many countries like Germany, Britain, France, and the United States downplayed the severity of the outbreak. One of the only countries that was open about the pandemic was Spain, which is why the disease was known as the Spanish Influenza, (although Spaniards referred to it as The Purple Death.) The strain of the flu was so powerful, it killed up to 100 million people worldwide over a period of two years.

Although the flu virus can live on human skin for only five minutes, it can last for 17 days if it is on paper.

Although medication is far better at treating the flu nowadays, it still kills about 26,000 people per year in the US alone. The flu can become fatal if it develops into pneumonia or congestive heart failure.

118. Foreign Accent Syndrome
AKA Accent Amnesia
When people suffer amnesia, they can forget what accent they speak in.

This disorder is so rare, there is only one case reported every year since 1941.
Example - Ashley Manetta started speaking in an Australian accent after she suffered memory-loss, even though she was from Pennsylvania. She spoke in

the accent because she visited an Australian island several times throughout her life.

119. Fregoli Delusion
A Fregoli Delusion causes the sufferer to believe that several people are the same person in disguise. This delusion usually develops from a brain injury or a lesion in the temporal lobe. Antipsychotic drugs are recommended for Fregoil Delusion.

120. Gallstones
AKA Cholelithiasis
Gallstones are stones composed of bile, which is made in the liver. Once the stone is solidified, the bile can't absorb it, causing it to build in size and thus, impairing the body's ability to produce, store, and transport bile. Without bile, the body can't digest food, causing the person to become very ill. Since the bile has nowhere to go, it's likely that the person will vomit it up.

Gallstones are sneaky as they can build for five years before the person starts to show visible symptoms. The first sign is usually biliary colic. This excruciating pain is often felt in the right upper abdomen or below the rips. It usually last 15 minutes but can go on as long as five hours.

A person can develop gallstones due to genetics, losing or gaining weight quickly, or by having a gastric band.

The stones can be as small as a grain of sand or as large as a golf ball.

Although the gallstones can be removed, it won't stop new ones from forming. The only way to get rid of them is to surgically remove the gall bladder. It needs to be treated immediately as it can develop into pancreatitis or cancer.

Although the body can function without a gallbladder, the person will have difficulty breaking down fatty foods like ice-cream and French fries.

121. Gaming Disorder

A gaming disorder is a condition that causes a person to be so obsessed with video games, it affects their personal, social, occupational, and educational life. Although people have complained for decades that video games can be bad for players, it was only recognized as a condition in 2018 by the World Health Organization.

122.   Gigantism

If a tumor develops in the pituitary gland, it will produce an excess of growth hormones, which causes the person's size to rapidly increase.

Men are twice as likely to develop the disorder than women. There doesn't seem to be a specific cause for gigantism, and it seems to affect all races evenly.

If a person develops the tumor before they reach the end of puberty, they become extremely tall (7ft or taller.) This is called acromegalic gigantism. This can be treated by having the pituitary gland removed. Although hormone medication can slow down a giant's condition, it might take ten years for them to stop growing.

If a person develops the tumor after puberty, they tend to grow outwardly and have a very wide appearance. They may only be a bit tall (6ft 3) but they may have an overly large head or abnormally-sized hands. This type of gigantism is simply called acromegaly. Acromegaly is usually excruciating as it also enlarges the heart, internal organs, and thickens the skin. It causes the gums and jaws to widen, which spaces out the teeth, which is very painful.

Examples – The wrestler, Andre "The Giant" Rousimoff was 7ft 4 due to suffering acromegalic gigantism. As he got older, Andre developed acromegaly, causing his body to widen, especially his head. The disorder put incredible strain on Andre's heart and he died at the age of 46.

Although giants appear to have superstrength in films and comic books, most people who suffer gigantism are incredibly weak. At WrestleMania VI, Andre the Giant physically couldn't walk to the wrestling ring so the producers created a mini-ring for him to be transported on. To make him feel less self-conscious, almost every wrestler at the event entered the wrestling stage in a mini-ring just like Andre's. Although the mini-ring concept seemed to be devised to make the show more epic, it was simply to hide Andre's debilitating illness.

The tallest giant ever was Robert Wadlow, who was 8ft 11 when he died at the age of 22. Weirdly, his death wasn't caused by his height. Wadlow died from an infection in his foot caused by a parasite. Wadlow was still growing at the time of his death.

Angus MacAskill was the tallest person to ever live that didn't suffer from gigantism, standing 7ft 9. Despite being incredibly tall, MacAskill was perfectly proportioned and didn't suffer the horrific side effects that most giants sustain.

123. Glass Delusion

A glass delusion is when a person incorrectly believes something is make of glass e.g. a car, table, a person, or a specific organ. Bizarrely, this delusion was very common during the Middle Ages. People with this condition can act normal one moment and then be frozen in terror, believing they will shatter it they take a single step.

Example - Charles VI of France believed his bones were made of glass. Charles put iron rods in his clothes so he didn't "shatter" if he fell.

124. Gourmand Syndrome

This is a benign condition that causes a person to be obsessed with high-quality food. After a Swiss political journalist called Kevin Pearce suffered a stroke, he quit his job and became a food critic. Apparently, the stroke causes a lesion in the right hemisphere in his brain that caused him to only think, talk, and eat fine foods.

Example – Charles Boyle in the tv series, Brooklyn Nine-Nine.

125. Grammar Pedantry Syndrome

This is the clinical term for a person who is compelled to correct grammatical errors and spelling mistakes. Nowadays, they are known as Grammar Nazis.

126.   Graves' Disease
AKA Toxic Diffuse Goiter

This autoimmune disease affects the thyroid, causing muscle weakness, rapid heartbeat, weight loss, heat intolerance, and sleeping problems. However, the most distinguishable symptom is bulging and asymmetrical eyes. Antithyroid drugs, thyroidectomy, and radioiodine have proven to be effective treatments for Grave's Disease.

Examples – Sia, George HW Bush, Maggie Smith, Missy Elliot, Marty Feldman

127. Green Blood Syndrome
AKA Sulfhemoglobinemia

This disorder causes sulfur to bind to a hemoglobin that carries oxygen in the red blood cell, turning the blood green. The condition usually fixes itself over time but blood transfusions may be used in extreme cases.

128. Green Skin Condition
AKA Chlorosis, The Virgin's Disease

During the 18th century, women suffered a disorder which caused their joints to swell, stopped their periods, and gave them heart palpitations. Also, it turned their skin green.

If you thought green-skinned women sounded weird, it gets weirder. This disorder only seemed to affect the upper class. Also, since it stopped women's menstruation cycle, doctors assumed that chlorosis sufferers developed the ailment because they weren't having enough sex.

Bizarrely, this condition was caused by a lack of iron. That's right. The women were simply anemic. And it only took 200 years for doctors to figure this out.

This condition was mentioned in the play, Pericles, which was partially written by William Shakespeare.

129. Guillain-Barre Syndrome
AKA Landry's Paralysis, Postinfectious Polyneuritis
This disorder causes the immune system to attack the nervous system which leads to weakness, numbness, tingling, and paralysis. Although the numbness begins in the hands and feet, it can lead to paralysis in only a few hours. Immunotherapy and physical therapy are necessary to slow down the disorder.
Example - Although President Franklin D. Roosevelt was originally diagnosed with polio, recent doctors state that his symptoms seem to correlate with Guillain-Barre syndrome.

130. Gynecomastia
This endocrine system disorder causes a man to develop breasts. Gynecomastia can develop due to an increased estrogen/androgen ratio, cirrhosis, cancer, medication, or chronic kidney disease. If treated with aromatase inhibitors, the breasts should decrease in size within two years. If the gynecomastia isn't treated quickly, it can become a chronic disorder.
Example – Meat Loaf's character, Bob, in the film, Fight Club.

131. Hair-Pulling Disorder
AKA Trichotillomania, which means "pull madness"
This causes a person to pull out their hair, eyelashes, and eyebrows compulsively. It is more common in women and is usually linked with stress, OCD, and dysmorphia. The drug, clomipramine, proves to be effective against this disorder.

Example – Colin Farrell suffers this condition. He wears a hat to stop his compulsion to tear out his hair.

132.   Halitosis

You probably think I'm going to say that halitosis is a condition that gives you bad breath. But it isn't. In fact, halitosis isn't a real disorder. The Listerine company concocted this "illness" to sell their product.

Although this might sound like a bold claim, Listerine had the exact same ingredients when it was sold for its original purpose – floor cleaning.

Although bad breath can be causes by abnormalities with the sinuses, throat, lungs, stomach, or esophagus, it is not a disorder in itself.

133.   Hallucination

A hallucination is perceived sense in the absence of stimulus. Most people believe a hallucination is something that you see that is not there. However, people can smell, touch, taste, and hear things that aren't there. Since schizophrenics hear false sounds, that can be classified as an auditory hallucinogenic disorder.

If a person hallucinates after losing a loved one, it is called a Grief Hallucination. This is extremely common with the elderly.

Although hallucinations are often caused by a mental illness or trauma, they can develop from consuming a Brugmansia flower, or overdosing on certain chemicals including solanine and LSD.

Example – Bob Hoskins plays the lead role in the film, Who Framed Roger Rabbit! Since the story revolves around cartoons, Hoskins had to envision imaginary characters every day on set. This put such an effort on his imagination, he suffered hallucinations for months. It was so exhausting, Hoskins had to take a break from acting for a year.

134. Hansen's Disease

Hansen's is caused by bacteria called Mycobacterium leprae and Mycobacterium lepromatosis.

This illness damages the nerves in a person's body, hindering or disabling their ability to detect pain. Although not being able to feel pain in the hands and feet sounds amazing, it has dire consequences. A Hansen's sufferer will often lean against something sharp or scalding without noticing. This can lead to infections that are so bad, the person may need their toes or fingers amputated.

For centuries, Hansen's was mistaken for leprosy. But here's the thing – leprosy doesn't exist. Weirder still, it never has. Leprosy is perceived as a disease that causes the sufferer's fingers and toes to rot away until they drop off. There is no disorder where this is a primary symptom. Although Hansen's sufferers may lose fingers or toes, this is because of infections and not from the disease directly.

Society was so scared of catching "leprosy" for centuries that people were banished from their homes throughout Europe for the most basic skin abnormalities including psoriasis, eczema, and rashes.

Hansen's can be spread through a cough or contact with fluid from the sufferer of the disease. However, it is not as contagious as many people believe and most of society is immune to the disease.

Although Hansen's is nowhere near as common as it used to be, it still affects over 500,000 people per years. 50% of Hansen's cases occur in India.

Hansen's is treatable with multidrug therapy. After a person catches Hansen's, the disease may incubate for up to eight years before the person suffers major symptoms. If the victim has the Hansen's treated before the incubation is complete, they should make a full recovery.

There's a "random fact" on the Internet that armadillos and humans are the only animals that can develop Hansen's. Although this is true, it's not in the way you may think. Although armadillos can catch Hansen's, it works as a pathogen so they carry the disease but don't suffer any of its symptoms.

135. Happy Heart Syndrome
In extremely rare cases of overexcitement, the mid-ventricle of the heart can swell, causing a heart attack. This is most common during birthdays and weddings.

136. Hashimoto's Syndrome
This is an autoimmune disease that causes a person to need far more sleep than normal. Very little is known about the disorder and so, there is very little treatment for this condition.
Example - Kristen Devanna developed this disorder when she was only 21. If she doesn't sleep for at least 18 hours daily, Devanna's skin starts cracking and she will constantly feel cold. Without a sufficient amount of sleep, Devanna can't perform basic tasks like have a shower or making a bed.

137. Hay Fever

AKA Allergic Rhinitis, Summer Catarrh, Hay Asthma

Hay fever is an inflammation in the nose after the immune system is exposed to airborne pollen. It usually triggers asthma, dermatitis, or conjunctivitis.

It is known as hay fever since it was erroneously believed that the smell of hay during the summer triggered the condition in certain people.

People who aren't exposed to pets in their first year are more likely to get hay fever.

Kiwis, salads, and spices are effective at curbing hay fever but there is no cure.

138. Headache

A headache is a pain in the head or neck. It can be caused by a cold, head injury, stress, fasting, fatigue, sleep deprivation, thirst, or a loud noise.

Although it seems like the most well-known condition, you would be surprised to learn that there are over 200 types of headaches including thunderclap, hemicrania continua, and hypnic headaches. However, they are usually split into three categories –

i)   A tension headache is a mild but irritating pain affecting both sides of the head. It can be remedied easily with aspirin or ibuprofen.

ii)   A migraine is a severe headache that usually affects one side of the head. These headaches can last 72 hours. Migraines are more common with women. The pain can be so piercing, it can cause the person to vomit.

iii)   A cluster headache is one of the worst pains a human being can suffer. It is a pain so unbearable, it's impossible to do any activity. Although most headaches are felt in opposing sides of the head, a cluster headache concentrates on a single spot, usually behind the eyes. People who have suffered this pain describe it as if the inside of their skull is being stabbed and burned simultaneously. Cluster headaches can last from 15 minutes to three hours. They are so-called since sufferers experience the headaches in clusters for weeks at a time. These headaches are so severe, it's not uncommon for sufferers to commit suicide. They usually occur to alcoholics after they suddenly quit drinking.

139. Heart Attack
AKA Myocardial infarction

A heart attack is caused when flow to a part of the heart gets blocked, which can destroy some of the muscle. If the tissue is damaged, the heart may not be able to pump blood around the body.

Heart attacks are so common in the morning (especially on Monday,) physicians call this time "the witching hour."

Men and women have different symptoms before and during a heart attack. A woman may feel light-headed, anxious, fatigued, or have a bloated stomach prior to the myocardial infarction. A man might feel a toothache, a tightness in the chest, pain in the shoulder or jaw, or a cold sweat. Prior to a heart attack, a person may feel a chest pain since not enough blood is being pumped to the heart. This is known as angina, which is Latin for "chest strangle." The common symptom for both genders is pressure in the chest, shoulder, and arm. At the first sign of a heart attack, chew an uncoated aspirin to slow down the amount of damage to the heart muscles.

Although a person can recognize the symptoms of a heart attack and take precautions, the sad truth is that half of heart attack sufferers die within the hour.

Sometimes, there are no symptoms prior to a heart attack. This is known as the Silent Heart Attack and usually occurs to people over the age of 75.

Heart attacks can be caused by high cholesterol, diabetes, old age, bad diet, genetics, obesity, alcohol use, and high blood pressure. Smokers are three times more likely to suffer a heart attack.

140.   Heart Disease
AKA Cardiovascular Disease
"Heart disease" is an umbrella term since there many versions of it including congestive heart failure, hypertensive heart disease, venous thrombosis, and cardiac arrest. The most common type is coronary artery disease, which is caused when a coronary artery that supplies blood to the heart is narrowed or blocked.

Heart disease is the leading cause of death in men and women in the US.

Heart disease can develop due to smoking, diabetes, obesity, lack of exercise, high blood cholesterol, bad diet, excessive alcohol consumption, or genetics.

Scientists have analyzed autopsy reports and have worked out that 72% of males and 67% of women that die of heart disease have a diagonal crease in their earlobes.

It might sound silly but laughter is one of the best ways to avoid heart disease since it expands blood vessels, which helps protect the heart.

141. Heartburn
AKA Acid Indigestion

Heartburn is a burning sensation in the chest but it can occur in the neck, throat, or jaw. Heartburn arises when stomach acid builds up so much, it re-enters the esophagus. One odd but surprisingly effective remedy at combatting heartburn is chewing gum. Gum increases the production of saliva, which neutralizes the stomach acid.

Avoid berries, tomatoes, and citrus fruits as they exasperate heartburn.

142. Hemianopsia

Hemianopsia cuts a person's vision in half. The word "hemianopsia" is Greek for "half without seeing." The weird thing about this condition is a person may not notice they have developed hemianopsia at first. You see, this disorder doesn't blur your vision. You can have 20/20 vision but still suffer hemianopsia. Imagine driving your car but you can only see the left lane. If you turn to face the right, the left lane disappears from your vision and it's impossible to see both lanes at the time. There is an extreme version of this disorder called quadrantanopia, which impairs three-quarters of a person's field of vision. If this condition isn't treated quickly, it can be permanent.

143. Hepatitis

Hepatitis is an inflammation in the liver. Hepatitis leads to skin and eye discoloration, vomiting, fatigue, abdominal pain, and diarrhea. Long-term hepatitis can lead to liver failure or liver cancer.

A person can develop hepatitis from a virus, excessive alcohol consumption, medication, and autoimmune diseases.

There are five kinds of hepatitis –

i)   Hepatitis A and E are caused by contaminated food and water. Hepatitis A and E are quite tame and treatable.

ii)   Hepatitis B is usually sexually transmitted. There is a high chance that a person with Hepatitis B will make a recovered within six months. If they don't, the inflammation becomes chronic and the so, the person will need medication for the rest of their life.

iii) Hepatitis C develops from infected blood. It's likely that Hepatitis C will become chronic.

iv)   Hepatitis D can only infect people who have Hepatitis B. Patients become resistant to medication quickly so Hepatitis D becomes difficult to treat.

v)   Like Hepatitis A, Hepatitis E is normally caused by contaminated food and water. Hepatitis E usually resolves itself within a few months.

Examples – Pamela Anderson and Ken Watanabe both developed hepatitis C. Luckily, they both made a complete recovery.

Daredevil, Evel Knievel, developed hepatitis C after receiving a blood transfusion.

144. Hematohidrosis

Hematohidrosis causes a person to secrete blood. Although the term means "sweat blood," the bleeding can come from areas of the body without sweat glands. It is unknown what causes this disorder. Only a handful of people throughout history have had this condition.

145. Hemophilia

Hemophilia is a disease that stops a person's blood from clotting. This can cause a hemophiliac to bleed to death from a paper cut. Hemophilia used to be called The Royal Disease because European monarchs suffered it and passed it on genetically. Although there is treatment for hemophilia, there is no cure.

146. Hernia

A hernia is a hole in muscle tissue, which causes abdomen tissue to bulge through. Hernias occur when the body doesn't have enough collagen. Collagen is the protein that makes skin and muscles flexible. Collagen can break down due to genetics, age, or smoking.

Despite what many believe, you can't get a hernia from straining yourself. Although overexerting yourself is the easiest way to learn you have a hernia, it is not the root cause.

The most common type of hernia is an inguinal hernia. This is when the groin pokes through the muscle tissue.

If a hernia occurs when a person is born, it is known as a congenital hernia.

An external device called a truss is used to adjust the organ back into its original position. Surgery is needed depending on the organ or the severity of the hernia. A hernia requires immediate medical attention as it can develop into inflammation, obstruction, or a hemorrhage. Hernia surgery has been performed for over 3,500 years. It was performed on the pharaoh, Ramses V.

147.   Hikikomori
This is disease that mainly occurs to Japanese male youths. Hikikomori, which means "pulling in," causes the sufferer to withdraw from society.

Sometimes, the sufferer will not leave their bedroom for months and develop OCD-like habits like meticulously cleaning their bed or combing their hair for hours on end. Self-harm is a common with hikikomori sufferers.

A study funded by the Japanese government in 2010 calculated that 236,000 people in the country suffered from hikikomori.

148. Histrionic Personality Disorder
HPD is a personality disorder that compels the affected person to be the center of attention. The affected person tends to act seductively and exaggerates their emotional state. People who interrupt public events naked often suffer from HPD.

Although medication isn't effective, psychotherapy and cognitive therapy worked against HPD.

149.   HIV/AIDS
Human Immunodeficiency Virus Infection and Acquired Immune Deficiency Syndrome are caused by the HIV virus.

An infected person will develop flu-like symptoms 2-4 weeks after being exposed to the virus.

The next stage, clinical latency, is the transition from HIV to AIDS. Although this tends to last eight years, it can be between 3-20 years. Although there are few symptoms during clinical latency, the patient will suffer weight loss, muscle pains, and fever at the end of this stage.

To understand how this virus works, I need to explain lymphocytes. A lymphocyte (or T-cell) is a white blood cell that is in charge of maintaining immunity in other cells. A healthy human being has a T-cell count of 500-1600. When the virus causes the T-cell count to drop below 200, that person has reached the final stage - AIDS. The immune system will be so low during this stage, the person is prone to diarrhea, sweats, fever, weakness, swollen lymph nodes, and chills. In the final stages, they may suffer meningitis, retinitis, pneumonia, tuberculosis, and tumors in the lungs, skin, and stomach.

The virus is passed by being exposed to infected blood or breast milk. A person can develop HIV by having sex with a person infected with the disease. HIV can survive on a dried blood clot at room temperature for six days. It can survive for weeks if the blood is in a syringe.

The original viruses, HIV 1 and HIV 2, are believed to have originated in primates in West-central Africa. African green monkeys in Cameroon infected with the Simian Immunodeficiency Virus (SIV) spread the disease to chimpanzees, which mutated into HIV 1. A bushmeat hunter was probably bitten by an infected primate while hunting, allowing the virus to infect him. Although this can't be irrefutably proven, it is accepted by the scientific community and is known as The Bushmeat Theory. As the virus spread from human to human, it mutated into what is now known as HIV. The first person recorded with HIV was a man from the Democratic Republic of Congo in 1959.

In May 1969, an African-American teenager called Robert Rayford became the first person to die from the disease. It was unknown for ten years what killed him.

When Gaetan Dugas died from AIDS in 1984, it was confirmed that HIV was the cause of AIDS. Since Dugas was gay, the disease became associated with homosexuality. In fact, AIDS was originally known as GRID (Gay-Related Immunodeficiency.) Although many people still believe homosexuals are the most common people to contract AIDS, this is untrue. The most common carriers of AIDS are heterosexual women between the ages of 18-25.

It was also believed that the virus could only be passed by having sex with an infected person. Because of this, AIDS was perceived as a "dirty" disease. When a heterosexual tennis player called Arthur Ashe died from AIDS after undergoing a blood transfusion, it became common knowledge that the virus was not only passed through sexual intercourse nor was it only killing homosexuals.

Thanks to advances in medicine, the average person infected with AIDS will live to 69 years. In fact, if you get treatment shortly after developing HIV, you will not necessarily develop AIDS.

10% of Europeans are immune to HIV due to

having an ancestor who survived the bubonic plague.

HIV and AIDS can be treated with antiviral therapy, herbs, cannabis, and maintaining a strict diet. *Examples* – Magic Johnson, Rock Hudson, Kevin Peter Hall, Anthony Perkins, Gia Carangi, Isaac Asimov, Charlie Sheen, Liberace, Freddie Mercury.

150. Hives

Hives is like a rash except the red markings have a raised, bumpy texture. Hives usually have a stinging burning sensation. Ancient Greek physician, Hippocrates, called hives "urticaria," which means "nettle."

Hives are often triggered by hay fever or asthma. Hives aren't too serious and should only last a few days. If they last longer than six weeks, consult a doctor for antihistamines.

Nettle tea, coconut oil, and chamomile tea have been proven to be effective against hives.

151. Hoarding Disorder

A hoarder has an inability to throw away seemingly unimportant material. Over time, a hoarder can have rooms filled with junk which is unused or broken. As the hoarder runs out of space, they will compromise by storing their material in bizarre places like a bathtub, fridge, or the attic. In situations like this, the hoarder won't be able to use the fridge or bathtub, which will cause them to become unclean and malnourished.

Although hoarding is a well-known concept, it was only classified as a disorder in 2013. Before that, it was classified as Obsessive-Compulsive Disorder since

18% of people with OCD are hoarders.

Different hoarders collect material for different reasons. One person may collect hundreds of canned food, believing that a nuclear fallout is inevitable. A has-been celebrity may keep every newspaper that mentions them to remind themselves that they were once famous.

Hoarders are usually considerate and will collect things that they believe to be useful to others.

Cognitive therapy tends to be more helpful with hoarding than medication.

152. Housemaid's Knee
AKA Prepatellar Bursitis
Housemaid's Knee is an inflammation on the knee. It's usually caused by an injury to the leg or by pressing on the knee excessively. It got its name from British servants who suffered inflamed knees due to performing housework excessively.

153. Hunchback
AKA Kyphosis, Roundback
A hunch is an excessive convex curvature of the spine. This causes back pain and stiffness. When people hear the word "hunchback," they probably think of an exaggerated hunch like the one Quasimodo has in Victor Hugo's story, The Hunchback of Notre Dame. However, mild hunches are incredibly common, especially in the elderly. A hunch can develop due to a disease, an injury, or through bad posture. Sometimes, a hunchback has to wear a brace to correct their curvature. If the curve is greater than 70 degrees, a doctor will probably recommend surgery.

154. Huntington's Disease
AKA Huntington's Chorea
Huntington's is a type of dementia that causes brain cells to die. HD effects 0.01% of people. At first, this form of dementia usually causes mood swings or motor functioning problems. As the disorder worsens, the person makes involuntary jerky movements and will have little to no control over their body. As the facial muscles stop working, the person will have difficulty talking or swallowing. In fact, most Huntingdon's sufferers die from choking.

A person with this active gene has a 50% chance of passing it onto their children. Although the disorder doesn't normally activate until the person is in their 30s or 40s, it is possible to tell if a fetus will develop Huntingdon's while it is still in the womb by taking a sample from the placenta cells.

According to the FDA, Tetrabenazine and Auestedo are the most effective drugs at slowing down Huntingon's. Sadly, there is no cure for this disease.

155. Hyperhidrosis
Hyperhidrosis causes a person to sweat profusely. Sometimes, a person can sweat to death. It is caused by an overactive sympathetic nervous system. It can be remedied with surgery or electrical treatment.

156. Hypohidrotic Ectodermal Dysplasia
AKA Anhidrotic Ectodermal Dysplasia, Christ-Siemens-Touraine Syndrome
This disorder impairs a person's ability to grow hair, nails, teeth, or sweat glands. Because the person

cannot sweat, they can die if exposed to intense heat. Sadly, there is no cure.

Example – When Wes Craven was casting the film, The Hills Have Eyes, he wanted to find the oddest person possible to play the cannibalistic mutant, Pluto. In the end, Craven cast Michael Berryman, who was born with HED. Since Berryman stands 6ft 2 and has no hair, fingernails, or sweat glands, he usually plays maniacs, monsters, demons, or aliens in films.

157. Hyperinflation Disorder

Okay, it's time for a history lesson. Between June 1921 to January 1924, the Weimar Republic of Germany suffered from hyperinflation. The German citizens owed so much debt after World War I, bank notes and coins became worthless.

The government thought they could counter this by simply printing more money but the debt skyrocketed to an absurd level. At the time, the currency was the mark. In 1922, one dollar was the equivalent of 320 marks. One year later, a dollar was the equivalent of a trillion marks!

Imagine if you bought a cup of coffee at a shop for $1. After you finish drinking it, you decide to have another. But in that time, the price of coffee has gone from $1 to a $1 million. Money became so worthless, the German inhabitants used it as wallpaper or gave it to their children to play with.

Now you might think – Wait, this isn't a disorder. It's just a bizarre situation.

But the hyperinflation had a psychological effect on people. The citizens of Weimar could only think of one thing – numbers. In case you were wondering, a trillion looks like this – 1,000,000,000,000. Imagine you had that amount of money in your pocket (somehow.) Now imagine that's not enough to buy a loaf of bread or toilet roll. And the hyperinflation didn't dip at all until 1924. This means it consistently escalated for two-and-a-half years.

I want you to picture 13 people.

Now picture 19 people.

Now picture 23 people.

See how tricky that is? Now picture a billion people. Now picture a trillion people.

The human brain was never designed for numbers of this magnitude. This caused people to develop a disorder where they said they had five trillion kids or two billion cats. Their minds just snapped as they couldn't process the numbers they had to deal with every moment of every day for nearly three years.

For decades, this disorder was known for only happening in one location in one time period. However, it happened again in Zimbabwe between 2007-2009 when prices skyrocketed to governmental incompetence. During this time, a trillion dollars couldn't buy a quarter-loaf of bread.

158. Hypersomnia
AKA Hypersomnolence

This disorder is like a combination of narcolepsy and insomnia. Now, this might sound a bit confusing. Narcolepsy causes the sufferer to go to sleep involuntarily. Insomnia prevents the person from going to sleep. So how can you have both conditions at the same time?

A hypersomniac usually cannot go to sleep

voluntarily. Normally, they can only rest when the narcolepsy kicks in. A hypersomniac can stay awake for days no matter how drowsy they feel.

Example - Freedom fighter and abolitionist, Harriet Tubman, developed this disorder after being hit on the head as a child. Tubman suffered the disorder for 50 years before having it corrected with surgery.

159. Hypochondria

A hypochondriac is constantly terrified they will get ill or they think they are ill when they are perfectly healthy. A hypochondriac can be genuinely ill but they might exaggerate their symptoms.

Hypochondria is a form of attention-seeking. This is clear when a hypochondriac seems disappointed when a doctor tells them they are healthy. To convince themselves they are truly sick, a hypochondriac may develop paranoia and believe everyone is lying to them. The hypochondriac may switch doctors, believing their doctor is "in on it."

Example – Charles Darwin kept a notebook for his health, which he updated every day.

Adolf Hitler always travelled with a doctor, terrified that he would get sick.

Writer, Hans Christian Anderson, wouldn't eat meat, worried that there could be a nail in it.

Florence Nightingale was so certain she was ill, she spent more of her life in bed than out of it. Nightingale was nearly completely bedridden for the last 57 years of life because of an illness she didn't even have.

160. Hypothermia

A human's average body temperature is 37.1 degrees

Celsius (98.6 degrees Fahrenheit.) When a person's body core temperature is below 35 degrees Celsius (95 degrees Fahrenheit,) body-heat dissolves faster than it can absorb, which develops into hypothermia.

Hypothermia causes a person to shiver, become dizzy, and have impaired motor functions. Hypothermia causes such intense damage to the nerves, the person may feel boiling hot to the point where they will undress, even if they are in a subzero environment.

An elderly person is more likely to develop hypothermia due to their weak immune system.

There is a misconception that drinking alcohol is effective at combating the cold. However, alcohol makes you colder, but you can't feel it because the drink causes blood vessels under the skin to dilate, bringing more blood to the surface. This gives the false impression that you are getting warmer.

If someone develops hypothermia, it's vital to warm them with blankets as soon as possible. When an elderly person develops hypothermia, it's likely they will be unresponsive to food. In this case, it's important to give them warmed intravenous fluid. If the hypothermia is critical, the person will need to be resuscitated or be given oxygen.

161. Hysteria

Although referring to someone as "hysterical" is an expression, it was incorrectly regarded as a mental illness for millennia. Ancient Greeks believed a woman's womb could wander around her body, which made her lose her senses. The word "hysteria" is derived from the Latin word "hystericus," which means "of the womb." Also, "uterus" means "wandering womb." The "symptoms" of this supposed disorder were insomnia, fainting, anxiety, and a "tendency to cause trouble."

Although the idea that the womb moved around the body was dismissed after a few centuries, hysteria was a recognized disorder until the 1950s.

162. Imposter Syndrome
AKA Impostorism, Fraud Syndrome,
This is a psychological disorder where a person doubts their own accomplishments and becomes convinced they will be revealed as a fraud.

163. Insomnia
Insomnia is an inability to sleep. Primary Insomnia is caused from stress, anxiety, environmental factors, or depression. Secondary Insomnia is caused by a medical condition.

Due to not having enough sleep, the insomniac can suffer memory problems, irritation, and aggression.

There are many types of insomnia –
i)   Transient Insomnia can last for a few days or weeks. It tends to happen when a person is suddenly stressed due to a new factor in their life.
ii)   Acute Insomnia can last for up to six months.
iii) Chronic Insomnia lasts for years. This can cause so much damage to the neurons, the person can develop depression or anxiety.
iv)   Fatal Familial Insomnia is a disease that eats away at the thalamus of the brain, which control's the person's sleep cycle. This can

render the person incapable of sleeping for 6-18 months due to relentless anxiety attacks. Since there is no cure, many sufferers will die from the disease.

There are dozens of techniques to help insomnia. However, every insomniac is different so it's important to try as many potential remedies as possible. If one technique doesn't work, try another. Medication, herbs, and therapy are helpful at combating insomnia.

Although it's a cliché, having milk and cookies in the evening is effective at making a person feel sleepy. However, you need to avoid large meals before bedtime.

If you find yourself tossing and turning, switch to another bed.

You can have a shower just before bedtime since it causes the body temperature to drop, which can trigger drowsiness.

If you can't sleep, the worst thing to do is lie in bed. If you are struggling to sleep, reading a book for five to ten minutes by a dim light should help you doze off. (Do not read on a computer as the screen's light can make you feel more awake.) It's also important to dim the lights two hours before bedtime.

164. Intermetamorphosis

This is a delusional misidentification syndrome that convinces a person that two people's minds have swapped bodies. The sufferer may believe that they saw someone morph into another person or entity. This disorder is very common with people who suffer face-blindness and Alzheimer's.

165. Intermittent Explosive Disorder
IED is a condition marked by random fits of rage which is disproportionate to the current situation. The person usually feels deep remorse afterward.

Psychotropic medication or cognitive behavioral therapy is recommended.
Example - Donald Duck

166. Internet Addiction Disorder
AKA Problematic Internet Use
A person with this disorder uses the Internet excessively. (Facebook Addiction Disorder falls under this bracket.) Although some people use the Internet for practical reasons, it is considered an addiction when it affects the person's well-being. A person with IAD will neglect to eat or sleep. It can be so bad, the person may not get up to go to the toilet.

Time management exercises and pharmacologic therapy are helpful remedies. In extreme situations, the addict may need electrical nerve stimulation

167. Iron Overload
AKA Hemochromatosis
This condition causes a person to produce more iron than their body can handle, which can develop into diabetes, heart failure, or liver disease. This disorder is so common in Ireland, it is nicknamed The Celtic Curse.

168. Irritable Bowel Syndrome
AKA Spastic Colon, Mucous Colitis
IBS produces abdominal pains, which changes the

pattern of bowl movements. This can cause diarrhea or constipation. Although it's unclear what causes IBS, it affects 12.5% of people in the developed world, 66% of which are women. Weirdly, the disorder is not caused by any disease in the bowel, which makes it very hard to diagnose.

IBS can be curbed by avoiding alcohol, red meat, fried food, dairy, caffeine, and sweeteners.

169. Jargon Aphasia

You know how children talk in gobbledygook sometimes? Well, some adults do that too. But they don't speak this way to be cute. They talk like this because they suffer jargon aphasia. They might substitute a specific word, a random word, or an entire conversation with gibberish.

Worse still, it's impossible to learn an aphasian's made-up words because they keep changing. An aphasian might call an egg an "enrodian gruster" one time and a "32B8" another time.

This disorder also effects the sufferer when they write so they will have incredible difficulty communicating.

Jargon aphasia is usually caused by a stroke, a brain tumor, a head injury, Parkinson's, or Alzheimer's.

170. Jerusalem Syndrome

This is a mental disorder that causes the sufferer to believe they are a Biblical figure such as Noah, Solomon, Mary, Moses, or Jesus. Sometimes, the sufferer doesn't think they are a Biblical figure but they are connected to one. A girl called Haley appeared on Dr. Phil claiming that she was about to give birth to Jesus Christ. Haley insisted she was giving birth to the Messiah since she had piled on weight for nine months despite never having sex. It turns out that she wasn't pregnant, and her weight gain was due to trapped wind.

You may wonder, "What if several people who believed they were Jesus Christ met each other?" Interestingly, this concept was tested by Milton Rokeach. Three patients in his psychiatric institution believed they were all Jesus Christ. As a test, Rokeach put the three patients in the same room to see what happened. One of the patients believed the other two were the reincarnation of prophets but not Jesus. One of the other patients thought the situation was a test of faith to see if he would doubt his own divinity. The last of the three simply said that the other two weren't the reincarnation of Christ and they were simply crazy. This scenario inspired a film in 2017 called Three Christs.

It is worth mentioning that reincarnation is not a part of the Christian belief.

171. Jet Lag
AKA Desynchronosis Circadian Dysrhythmia
Jet lag is when a person has difficulty adjusting their sleep cycle to a new time zone after travelling via plane from one side of the world to the other. It usually takes one day per time zone to adjust e.g. If you travelled from New York to London, which has a five-hour difference, it will take five days to adjust.

Jet lag causes more effects to your body than you would imagine including your body temperature and melatonin production. Since pineapple has an abundance of melatonin, eating this fruit should return your sleep-pattern back to normal. The low oxygen on a plane can make a passenger feel faint so it's important to stay as hydrated as possible to combat jet lag.

172. Joker Syndrome
AKA Witzelsucht

A person with Joker Syndrome is compelled to say and perform inappropriate things while being oblivious to how inappropriate they are.

This disorder is normally caused by frontal lobe damage or a tumor on the right frontal lobe.

The venlafaxine drug has proven to be incredibly effective against Joker Syndrome.

Example – The Batman villain, the Joker has Joker Syndrome. Obviously.

Also, James Bond has this condition since he makes puns immediately after killing henchmen.

173. Kidney Stones
AKA Urolithiasis

This disease causes stones to form in a person's urinary tract. The stones are so small, they usually pass from the kidney and out of the body without any consequences. If the stone is over five millimeters, it can cause a blockage in the ureter, which causes incredible pain in the lower back.

Kidney stones can be massive. A six-inch wide stone was removed from a man in 2011.

Although 10% of people get kidney stones at some point, men are twice as likely to develop them.

Although it is believed that these stones are made of calcium, this isn't true. They are formed by salt. Kidney stones can form if a person doesn't drink enough water and consumes a lot of processed foods and deli meats.

If you have had kidney stones, it is likely it will return within the next ten years.

90% of kidney stones are passed through the urine. If this occurs, it's vital to take the stone to a doctor so they can analyze it to figure out how to avoid another one developing in the future.

If the kidney stone cannot be passed naturally, a doctor can use a lithotripsy machine that shatters the stones with shockwaves until they are small enough to pass.

If you don't want to go to a doctor, there is an alternative solution – ride a rollercoaster. That's not a joke. According to urologist, David Wartinger, riding at the back of a rollercoaster while it plummets downward is astoundingly effective at letting a person pass kidney stones. The Big Thunder Mountain rollercoaster in Walt Disney World, Orlando is The Kidney Stone-Passing Capital of the world... which is a pretty weird thing to be proud of.

174. Kleptomania

Kleptomaniac is an impulse control disorder that compels a person to steal. Kleptomaniacs usually don't steal to save money. Instead, they steal for control, revenge, or to get a buzz. According to one study, 0.6% of Americans are kleptomaniacs. In this same study, 65% of kleptomaniacs are bulimic. Women are more likely to suffer from this disorder.

Cognitive therapy has shown that kleptomania is usually a side effect from another disorder like anxiety, depression, or hoarding.

175. Klippel-Trenaunay Syndrome
AKA Hemangiectatic Hypertrophy
This birth defect causes a person to develop varicose veins, a port-wine stain, and an underdeveloped lymph system. Some people with KTS suffer abnormal bone and soft tissue growth, which cause a specific limb to increase in size and strength.
Example – Matthias Schlitte is a professional armwrestler who suffers KTS. Since he weighs 150lbs, he qualifies as a lightweight arm-wrestler. Since his disease causes his arm to be 33% larger than normal, he's practically unbeatable in armwrestling.

176. Koro
Koro is a mental disorder that convinces the sufferer that his genitals are shrinking or disappearing.

177. Kuru
During the 1950s, there was a record of a Papua New Guinean tribe seemingly laughing and shaking to death. This disorder, kuru, which means "to shake" in the Fore language, is caused by eating human brains. The cure is pretty simple - stop eating human brains.

178. Laron Syndrome
This syndrome mainly has benefits. (We need more syndromes like that.) Although this disorder causes dwarfism, it gives the sufferer an astounding immunity to diabetes and all forms of cancer. Many

people diagnosed with Laron Syndrome live for over 90 years. For some reason, Laron Syndrome is very common in Ecuador.

179. Laughter-Induced Syncope

This condition forces a person to faint if they laugh. It became more well-known after reports of a woman who fainted every time she watched the comedy series, Seinfeld. As a result, this condition is nicknamed Seinfeld Syncope.

180. Leukemia

Although leukemia is a cancer, it doesn't originate in a tumor. Instead, it originates in the blood, bone marrow, digestive tract, and lymphatic system. As the cancer cells spread, the marrow and blood cells become impaired.

The main symptoms of leukemia include spotty skin, swollen and bleeding gums, night sweats, joint pain, and tender lymph nodes.

To understand leukemia, I need to explain how blood works. White blood cells fights infections. The word "leukemia" is derived from the Greek for "white blood." If these cells don't form properly, the body can't fight diseases and viruses.

Red blood cells carry oxygen around the body. With the red blood cells unable to carry oxygen, the person will become pale and fatigued.

Leukemia also damages platelets. A platelet is a component of blood that allows it to clot. If the platelets stop functioning, the person won't be able to heal from bleeding or bruising.

Since leukemia also attacks marrow cells, it makes bones more fragile, which means they break easily and take much longer to heal.

Leukemia can be broken down into four categories –

i) Acute lymphocytic leukemia usually occurs in children (especially Hispanics.) This is the cancer that most children die from.

ii) Acute myeloid leukemia affects children and adults.

iii) Chronic lymphocytic leukemia mainly effects adults over the age of 55. Although it does affect younger adults, it never affects children.

iv) Chronic myeloid leukemia mainly affects adults but some children develop this form of cancer.

Sadly, the cause of leukemia is unknown. As a result, there is nothing that can be done to prevent leukemia.

Like normal cancer, treatment for leukemia includes chemotherapy, radiotherapy, and targeted therapy. A bone marrow transplant also has a high success rate. 54% of people with leukemia survive beyond five years. Once they reach the five-year mark, there's a good chance the cancer will not return.

181. Lisp

A lisp is a speech disorder that causes a person to thrust their tongue beyond their front teeth while speaking, which makes words with the letters "s," "c," "z," and "ts" come out inarticulately. Speech therapy is very effective at correcting lisps. If speech therapy doesn't work, a dentist can treat the lisp with a laser incision called a frenectomy. The procedure takes up to 15 minutes to correct.

182. Locked-In Syndrome
AKA Pseudocoma

Locked-In Syndrome prevents a person from moving any part of the body except the eyes. It can be caused by a stroke, brain trauma, poisoning, or a drug overdose. It can take up to three months before a person is correctly diagnosed since some doctors may believe the patient is in a coma.

Although this disorder is usually permanent, there are exceptions. Martin Pistorius developed LIS when he was 12. After eight years, Pistorius recovered. How, you may ask?

Three words – Barney and Friends. While Pistorius was being cared for in a hospital, Barney and Friends regularly played on his television. Pistorius found the theme-song so annoying, he "willed" himself to walk so he could turn off the television when the show came on.

Pistorius isn't overreacting. The Barney theme-song is considered to be so repetitive, it has been played to prisoners in Guantanamo Bay as a form of torture and no, I am not joking.

183. Lockjaw
AKA Tetanus, Grinning Death

Lockjaw is an infection which causes muscle spasms that are so severe, they can cause bone fractures. Because of the name, many people assume this disorder only occurs in the jaw. Although the infection begins in the jaw, it works its way down until it infects the rest of the body's muscles. The spasms usually trigger 3-21 days after the person is exposed to the infection. 10% of lockjaw sufferers die from the illness.

Lockjaw can be easily prevented with a vaccine. If lockjaw develops in a person before they have been vaccinated, they will need diazepam or another immunity medication. If the lockjaw is severe, the person may need to receive a tracheotomy since the muscle spasms can block the windpipe.

184. Lupus
AKA Systemic Lupus Erythematosus

This autoimmune disease causes the body's immune system to mistakenly attack healthy tissue throughout the body. This leads to mouth ulcers, fatigue, swollen lymph nodes, fever, hair loss, and chest pain. The most identifiable trait is a bright red rash across the face. Once a person develops lupus, he or she will have it for the rest of their life.

The word "lupus" is derived from the Latin for "wolf" since the rash is said to resemble a wolf bite (although many people believe the rash resembles the shape of a butterfly.)

Although it is unclear what causes lupus, genetics is a factor.

Although 90% of lupus sufferers are women, the disease is more severe in men. Lupus normally affects people between the ages of 15-45.

Treating lupus can be very expensive. When new symptoms develop, the sufferer can't just go to a doctor. It's more likely they will need to visit multiple specialists. If the lupus starts to affect the kidneys, the person will need to be treated by a nephrologist. If the lupus starts to damage the skin, the person will need to consult a dermatologist. Steroids like predisone are used to treat lupus but they have side effects such as depression, weight gain, and mood swings. In some cases, chemotherapy is an effective treatment.

Example – Claudette in the tv series, The Shield

185. Lyme Disease

This is an infectious disease caused by the Borrelia bacteria, which is spread by ticks. A week after receiving a bite from an infected tick, the person will suffer fatigue, fever, and headaches. In extreme cases, the person can develop facial nerve paralysis, meningitis, and arthritis.

Lyme Disease can be diagnosed with a blood test. For the first few months of developing Lyme disease, most tests will come out negative. Because most doctors can't recognize that a patient has the early stages of Lyme Disease, it's extremely difficult to treat. By the time a doctor can confirm that a patient does have Lyme disease, the illness will have exasperated, making it far harder to medicate.

Lyme disease can be cured if treated early. Otherwise, the person will need antibiotics for the rest of their life.

Examples – When singer, Avril Lavigne, developed Lyme disease, doctors dismissed her symptoms. She didn't get correctly diagnosed for eight months.

Real Housewives star, Yolanda Foster, can no longer watch tv, read, or write after developing Lyme disease.

186. Mad Cow Disease

AKA Bovine Spongiform Encephalopathy

This disease causes the brain and spine of a cow to shrink. If a person eats meat from an MCD-infected cow, that person can develop Creutzfeldt-Jakob Disease. CJD causes a person's brain to shrink at an alarming rate. There is no cure for the disease.

Weirdly, some Papua new Guineans are immune to Mad Cow Disease.

187. Maple Syrup Urine Disease

MSUD is a potentially fatal illness that breaks down amino acids throughout the body. Luckily, this disease gives the sufferer a huge warning; their earwax and urine will smell exactly like maple syrup.

188. Malaria

Malaria is an infectious disease caused by a mosquito bite. When a mosquito drinks a person's blood, the insect can pass on a plasmodium parasite that infects the body. Within two weeks of being exposed, the infected person may develop a fever, headaches, and fatigue. In extreme cases, the person can have seizures, fall into a coma, or die.

Although it's common knowledge that mosquitoes make a buzzing sound, only males do this. Sadly, it's the silent female mosquitoes that drink human blood.

Malaria has caused more human deaths than anything in history. Literally billions of people throughout history have died from this disease. Although malaria has been acknowledged for millennia, no one had a clue that it was caused by mosquitoes… until the 1930s! Before then, it was believed that malaria was caused by "bad air" during the night. "Malaria" literally translates into "bad air." Now that humanity understands how malaria works, doctors can treat it with artemisinins and other antimalarial medicine. However, the disease is still devastating and kills about 500,000 people annually.

If a person recovers from malaria, they will have a mild immunity from the disease for several years.

70% of deaths connected to malaria are in children younger than five. In some African countries, children can develop malaria six times a year. Also, pregnant women are more vulnerable to malaria.

It's possible to catch malaria from another person if you share needles or through a blood transfusion.

Weirdly, one of the only cures for this disease... is another disease – sickle cell. Sickle cell causes the red blood cells to become misshaped, meaning they are more likely to clot.

189. Marfan Syndrome

This disorder makes a person's bones grow too quickly. It overdevelops the bones in the arms and legs so they are disproportionally long. It also makes the bones overly dense so the person will have astounding strength.

Example - Abraham Lincoln suffered this disorder, which made him the tallest US president, standing 6ft 6. He was also the strongest president. While building cabins, Lincoln hauled crates of stones that weighed half a ton around his town. Out of the 300 wrestling matches he fought in, Lincoln only lost once.

Peter Mayhew, who is best known for playing Chewbacca in the Star Wars franchise, also suffers Marfan Syndrome.

190. Mary Hart seizures

A person with this disorder will suffer epileptic

seizure if they hear a specific tone and pitch.

The name comes from television host, Mary Hart, who hosted the program, Entertainment Tonight. In 1991, New England Journal of Medicine found a woman who suffered seizures if she heard Mary Hart's voice. This has been parodied in many television shows including Seinfeld, Animaniacs, and Tiny Toons Adventures.

191.   Measles

Measles is one of the most contagious infectious disease in the world. Symptoms usually develop 10-12 days after being exposed to the virus. At first, the person will have flu-like symptoms including a cough, eye inflammation, a runny nose, and a fever. The virus also causes the infected person to be covered in a full-body rash. The markings from the rash are known as Koplik spots and appear on the skin as well as the inside of the lips and gums. 33% of infected people develop another condition such as bronchitis or pneumonia. The infected person usually endures measles for 7-10 days.

Although measles usually clears after a week or so, there's a 0.1% chance it will spread to the brain. When this happens, doctors can't do anything to reverse it. The few sufferers who survive will most certainly go deaf or blind.

Measles is very rare in modern society thanks to vaccines. Before vaccines, practically every person developed measles before they were 15. Tragically, measles has made a comeback recently due to misinformation that vaccines cause autism. Measles caused 530,217 deaths in the US in 1900. In 2015, measles only caused 61 deaths. On June 28th 2018, it was announced that there was 52 cases of measles in the UK so far that year. The year before, there was one single case in the entire country. Every human being should be vaccinated as soon as possible to avoid running the risk of developing measles, mumps, whooping cough, or worse.

192. Megalomania
AKA Grandiose Delusions

Megalomaniacs believe they are far more important than they actually are. Sufferers of this disorder may erroneously believe they are wealthy, loved, famous, powerful, and immune to failure. A megalomaniac can delude themselves into thinking they are an important figure from history like Napoleon or Cleopatra. A megalomaniac may see themselves as magical or god-like. Many dictators and serial killers throughout history can be classified as megalomaniacs. The disorder is very common with schizophrenics and bipolar sufferers.

Example – The first emperor of China, Qin Shi, was so megalomaniacal, he referred to himself as a god and banned citizens from using the word "I" to refer to themselves as he considered it too respectful.

Gina in the tv series, Brooklyn Nine-Nine, depicts megalomania in a surprisingly realistic way. Despite routinely neglecting her duties as a police officer's secretary, Gina adamantly believes she's one of the most important people to ever live.

193. Melophobia
Melaphobia is a fear of music. Melaphobics don't just

dislike music. They hate it so much, it causes them to suffer long-term tinnitus.

194. Meningitis

Meningitis is an acute inflammation on the meninges. The meninges are a protective membrane that covers the spinal cord and the brain. The disease can be caused by fungi, bacteria, or a virus.

The first symptoms of meningitis are a severe headache and a stiff neck. Eventually, it can develop into falling blood pressure, rapid heart rate, and sepsis.

A person can die of meningitis less than a day after symptoms become noticeable. Although there are antibiotics for meningitis, 10% of those who are infected will die from the illness, even under the best care. Those who survive have a 20% chance of permanent brain damage. There are many other disorders that can manifest from meningitis including deafness and epilepsy.

Although meningitis tends to affect teenagers between the ages of 15-19, it can attack anyone of any age.

Luckily, there is a vaccine for meningitis. Treating the disorder depends on how it developed. Bacterial meningitis requires antibiotics and corticosteroids. Viral meningitis requires supportive therapy. Fungal meningitis requires antifungal medication.

195. Microcephaly

AKA Pinhead Syndrome

Microcephaly is a medical condition that causes a person to have a shrunken brain and head. This can happen before birth or in the first few years of a child's life. Microcephalics tend to suffer from poor motor function, impaired speech, abnormal facial features, dwarfism, seizures, and intellectual disability.

During Victorian times, microcephalics appeared in circus and were advertised as aliens.

Example – Schlittze in the 1932 film, Freaks, suffered this disorder. Although microcephalics don't usually live very long, Schlittze died when he was 70.

196. Mini strokes

AKA Transient Ischemic Attacks

Like a regular stroke, a mini-stroke causes a clot that blocks blood-flow in the brain. The difference between this and a normal stroke is a mini-stroke usually doesn't kill brain tissue. This should cause the person to develop weakness on one side of the body for a few minutes or hours.

Although it isn't as severe, you must consult a doctor immediately if you have a mini-stroke. According to a US study, at least 10% of people who suffered a mini-stroke experienced a stroke within three months. Sometimes, the stroke occurs less than one day after the mini-stroke.

Example - Although many sources state that Julius Caesar suffered from epilepsy, a 2015 study showed it was more likely he suffered from mini-strokes.

197. Molyneux's Problem

Molyneux's Problem is when a blind person suffers sensory overload after obtaining the ability to see. After Virgil Adamson was blind for 40 years, he had his cataracts removed, allowing him to see. However, Virgil felt terrified, physically ill, and depressed since his brain couldn't process objects, faces, movement, words, depth, color, etc. He had difficulty telling the difference between an apple and a photograph of an apple because he couldn't process the difference between a 2D object and a 3D object. Sadly, some people with this condition may never be able to process visual information the way a regularly sighted person can.

Example – The 1999 film, At First Sight, which is based on Oliver Sacks story, To See and Not See, shows how a blind man deals with the changes to his life when he gains the ability to see.

198. Motion-Blindness
AKA Akinetopsia

This condition makes a person incapable of perceiving motion. Imagine if you were watching a film but some frames were cut so the motions appeared jumpy. Motion-blindness is a condition that works like this every moment of every day. But you don't just miss a frame here and there. You miss up to 90% of every object moving in front of you. If you poured milk into a glass, you would see the liquid gradually fill up the container. A person with motion-blindness would see an empty glass one moment and then a see a full glass the next.

You can watch videos online that demonstrate how it feels to suffer this disorder.

The disorder can be caused by brain lesions or Alzheimer's disease. Sadly, there is no cure or treatment.

199. Multiple sclerosis

MS is an auto-immune disease that makes the body's defense against illnesses malfunction, causing the immune system to attack its own cells. Specifically, it attacks the myelin, which protects the nerves.

This causes many potential symptoms including fatigue, numbness, tingling, vision impairment, imbalance, mobility issues, and slurred speech. It also impairs a person's concentration, spatial ability, and problem-solving skills. Sometimes, the symptoms can come and go for several weeks or months. When the symptoms relapse, it is known as an exacerbation.

Although the cause of MS is unknown, scientists believe it is connected to a Vitamin D deficiency.

MS is nicknamed The Silent Disease as people with the illness look completely normal.

Although there is no cure, chemotherapy can effective at diminishing the effects of MS. Experts advise MS sufferers to avoid sun exposure, hot baths, or intense exercise.

Example – Selma Blair, Jack Osbourne, Richard Pryor

200. Mumps

Mumps is a highly contagious viral disease that causes one or both parotid salivary glands around the neck to swell. Other symptoms include muscle pain, headache, fatigue, fever, and lack of appetite. Mumps is spread from human to human by being exposed to an infected person's respiratory secretions e.g. a cough or a sneeze. The symptoms trigger 16-18 days after being exposed to the mumps virus. It takes 7-10 days for the mumps to run its course. Mumps is very rare in the developed world thanks to vaccines. If you never received the vaccine, you may wonder why you have never developed mumps. The weird thing is that you probably did and didn't notice. 33% of people who become infected with the mumps virus develop absolutely no symptoms. Like chickenpox, the disorder is more severe in adults than in children. Adults infected with the mumps virus can develop testicle inflammation, meningitis, deafness, infertility, and pancreatitis. Although women can develop ovarian damage, it doesn't always lead to infertility.

To relieve the symptoms of mumps, one should gargle saltwater, eat soft food, drink a large amount of water, and use heat-packs on sensitive areas.

201. Munchausen Syndrome
AKA Factitious Disorder Imposed On Self
Munchausen Syndrome is when a person pretends to be ill for sympathy, attention, assurance, or to shirk responsibility. The syndrome's name comes from the fictional character, Baron Munchausen, who was based on the 18th century storyteller, Karl Friedrich von Munchhausen. Since Karl was known for exaggerating his stories, another writer called Rudolf Erich Raspe parodied these tall tales in the 1785 book, Baron Munchausen's Narrative of his Marvelous Travels and Campaigns in Russia. This name was appropriated for Munchausen Syndrome as people with this "illness" usually concoct over-the-top symptoms.

There's a similar concept known as Munchausen Syndrome By Proxy (MSBP) or Factitious Disorder Imposed On Another. This is when a caretaker fakes or causes medical symptoms in a person they are caring for. The term was first used in 1977 by Dr. Roy Meadow when he suspected a patient's mother was poisoning her son with salt.

202. Mutism

A mute is a person who is unable to speak. It is usually caused by an injury, surgery, autism, undeveloped muscles, or a trauma.

i)   Selective Mutism is an anxiety disorder where a person chooses not to speak. Mutes prefer to communicate through sounds or gestures and tend to live in their own "bubble."

ii)   Alalia is the term used when a child has difficulty learning to speak. Albert Einstein fell into this category since he couldn't speak until he was four years old.

iii) Aphasia is when the cerebral center of the brain is damaged, rendering a person incapable of speaking

iv)   Aphonia is when a person can't speak due to suffering damage to the larynx

v)   Anarthia prevents the muscles in the mouth and jaw from developing, making it impossible for the person to speak.

vi)   Akinetic Mutism is the term used for a person who can't speak due to being catatonic.

203. Nail-biting
AKA Onychophagia

Nail-biting is a common form of impulse control. Most people bite their nails out of boredom or stress. Although it seems like a minor habit, nail-biting can cause damage to the nails, fingers, mouth, teeth, gums, and the digestive tract. Because the hands are the dirtiest part of the body, you can digest bacteria or pinworms if you constantly put your nails in your mouth.

Since most people bite their nails so they are relatively the same length, this can develop into Obsessive-Compulsive Disorder.

Behavioral therapy is a good way to treat nail-biting. Some people wear a bitter-flavored nail polish to discourage them from biting their nails.

204. Narcolepsy

Although narcolepsy is known as the neurological disorder that makes the sufferer spontaneously fall asleep, this is an oversimplification. A narcoleptic may not suddenly fall asleep but instead feel incredibly drowsy for seemingly no reason. Because of this, it's very easy to misinterpret a person's narcolepsy for fatigue or laziness. Only some narcoleptics fall asleep with little warning. When this happens, it's called a sleep attack and it can happen when the person is walking around, in the middle of a physical activity, or when they're driving.

You would assume that a narcoleptic will act very tired just before they have a sleep attack. However, some narcoleptics may become hyper-active, panicky, giddy, or aggressive just before they have a sleep attack.

The illness is caused by the loss of a brain chemical called hypocretin.

A dog can easily be trained to tell when a narcoleptic is going to have a sleep attack five minutes before it happens.

Although stimulants can be used to treat narcolepsy, there is no cure.

205. National Hotel Disease

Washington DC's National Hotel was a respected facility during the 1850s. But then something strange happened. People who stayed at the hotel started dying.

Now when a couple of people die in a specific location, it's common for that place to gain a reputation for "being haunted."

But the National Hotel was stranger than that. There weren't four or five deaths there. There were almost 40. Weirder still, they all died with the same symptoms – inflammation of the large intestine, diarrhea, swollen tongue, and constant vomiting. The hotel became so infamous that the disorder was known as National Hotel Disease. One of the people who suffered the condition was President James Buchanan. Luckily, he survived (despite suffering the condition twice.)

No one knows for certain what caused this condition but it was probably a type of dysentery caused by a burst pipe, spreading pollution into the water supply.

206. Necrophilia
AKA Thanatophilia, Necrolagnia

This condition causes the person to have an impulse to have sex with a dead person. As grotesque as that sounds, necrophilia isn't classified as a mental disorder according to The American Diagnostic and

Statistical Manual of Mental Disorders (DSM.) Necrophilia is classified as a paraphilia; a strange sexual attraction or practise between people or objects.

There are four categories of necrophilia –

i)   Homicidal – A person who kills the victim to fulfil the fantasy. 42% of necrophiliacs fall under this category.

ii) Regular

iii) Fantasizers – A person who finds the idea of necrophilia arousing but would never commit the act.

iv) Pseudo – A person who commits the act unplanned.

21% of necrophiliacs carry out the act for a "union with a lost love." Although 95% of necrophiliacs are men, there has never been a case of a female homicidal necrophiliac. Weirdly, only 15% of necrophiliacs admitted to carrying out the act purely out of sexual attraction.

207. Neotenic Complex Syndrome

NCT prevents a person from aging. When Brooke Greenberg was 20, people assumed she was a year old. Only seven people have been recorded with NCT. All of them were female.

208.   Night Terror

AKA Parasomnia, Sleep Terror, Pavor Nocturnus

Although many people believe they have suffered night terrors, they are extremely rare. A night terror isn't just a really bad nightmare. It is a dream of such intense dread, the person will be shaken for hours or days after awakening. Most people wake from a night terror with a blood-curdling scream while kicking and thrashing.

The night terror usually occurs in the first 3-4 hours of sleep.

Although dreams are usually fantastical, night terrors feel incredibly realistic. If you had a night terror that a murderer broke into your home, you might be convinced that is actually occuring upon awakening.

Like a regular dream, it is possible that the person will not remember the night terror at all. In fact, some people will have no memory of freaking out after awakening.

Night terrors tend to happen to people suffering from Post-Traumatic Stress Syndrome or a mental illness. They can also develop as a side effect from medication. Children normally grow out of night terrors. Adults usually require psychotherapy and diazepam to treat the disorder.

209. Nodding Syndrome

This disorder is a much more severe version of Bobble-Head Doll Syndrome. Not only does it happen to adults but it occurs even when the patient is trying to eat or sleep, which makes normal life impossible. It's common in Sudan, Tanzania, and Uganda. It's so common in Uganda, there are some villages where at least one person per family suffers the illness.

210.   Obsessive-Compulsive Disorder

People who suffer OCD develop strange obsessions or habits called rituals e.g. an inability to sit in blue chairs, the need to open and close every door three times before leaving a room, the need to clean a kitchen an incredibly specific way, etc. Despite what is depicted in films and television, many OCD sufferers do not have an obsession with being clean.

Bizarrely, it was only discovered in 2012 that people can develop OCD by developing a streptococcal infection. Stranger still, this infection can cause the person to develop Tourette's. If it develops from the streptococcal infection, it usually stops after a few months.

Example – Although shows like The Big Bang Theory or Monk have characters with OCD, experts say that the disorder is depicted inaccurately. Actress, Lena Dunham, has this disorder and plays an OCD sufferer, Hannah Horvath, in the show, Girls. Since she has had the disorder for years, Dunham depicts the character with incredible accuracy. Dunham says that most depictions of OCD on television are "too quirky" to be realistic. An OCD sufferer's rituals are much subtler in real life.

211. Ondine's Curse
AKA Congenital Hypoventilation Syndrome

OC scrambles the part of the brain that maintains unconscious breathing. Since a person suffering OC can't breathe instinctively, they have to consciously inhale every moment of every day. Sadly, it gets worse – a person with this condition can suffocate to death in their sleep.

Although most sufferers are born with this disorder, some people have developed it after an accident, an infection, sleeping pills, or even from alcoholism. The only way to survive Ondine's curse is with a breathing tube through the neck.

212.   Orthorexia
AKA Selective Eating
Orthorexia is an obsession with eating healthy food. According to the senior lecturer of the British Dietetic Association, Ursula Philpot, orthorexics are "solely concerned with the quality of the food they put in their bodies, refining and restricting their diets according to their personal understanding of which foods are truly 'pure'." An orthorexic's obsession is usually about feeling pure and balanced rather than being thin. They may focus on certain foods while neglecting certain food types. An orthorexic may limit how much carbohydrates they consume to lose weight. Although excessive carbohydrates can lead to weight gain, carbs are a basic need. If you cut down on them (or any food group,) it can lead to skin and nail damage, hair loss, vomiting, and anxiety.

Bizarrely, orthorexia is not recognized as an eating disorder in the Diagnostic and Statistical Manual of Mental Disorders. Even Wikipedia refers to orthorexia as a "proposed eating disorder."

213. Osteoporosis
Osteoporosis is a bone resorption disease. A normal person's bones are constantly remodeling. Although you can't feel it, your old bone cells keep being replaced with new cells. This process is so slow with a person with osteoporosis, it causes the bones to break easily. Also, the bones heal far slower than normal.

33% of women and 20% of men are prone to the

disease. Although the elderly is prone to osteoporosis, 2% of women in their 20s suffer the illness. The bones that are most likely to break are the vertebrae, the hip, and the forearm.

The disease tends to develop in people who have diabetes, menstrual problems, or poor nutrition. Soda weaken bones so much, some physicians nickname the drink as "osteoporosis in a can."

Although the disease has been recorded for millennia, it wasn't recognized until the 19th century. In 1820, physician, Jean Lobstein, coined the name "osteoporosis," which means "bone space."

Osteoporosis can be verified with an x-ray or a bone-density test. The disease can be treated with medication and Vitamin D. Regular exercise is important but it cannot be too intense. Despite what many people believe, calcium doesn't make bones stronger. Instead, it slows down the deterioration of marrow cells. Although consuming calcium is vital, it will not strengthen a person's bones.

Examples – Joan Rivers, Sally Field, Gwyneth Paltrow

214. Othello Syndrome
AKA Morbid Jealousy, Delusional Jealousy
Although we all get jealous, some people's envy is so intense, it is classified as a delusional disorder. A person with Othello Syndrome will accuse a partner of something with next to no evidence.

Symptoms of Othello Syndrome include –
i) Limiting or monitoring the time that their partner is outside
ii) Physically and verbally abusing their partner.
iii) Isolating their partner from their family and

friends.

iv) Going through their partner's belongings including their phone, messages, emails, and social media accounts.

v) Threatening to harm oneself.

vi) Denying their own paranoia

Othello Syndrome is nearly always egosyntonic. This means that the accusations are linked to the accuser's own insecurities. If a man is insecure about his body, he may accuse his partner of having an affair with every muscular man that she sees.

This type of jealousy can be caused many different ways – someone betrayed their trust, mental illness, trauma, etc. The key is to find the source of this jealousy so the person can get the correct treatment, be it counselling, couple therapy, family therapy, antidepressants, etc.

215.   Pain Immunity

There are several different disorders that render a person impervious to pain. This can be caused by a mutation in the protein, Zinc Finger Homeobox 2 (ZFHX2.) There is an Italian family composed of an elderly woman, her two daughters, and three grandchildren who have all been diagnosed with pain immunity. Sometimes, they feel intense injuries but it only lasts seconds. One of the girls, Letizia, broke her shoulder while skiing but didn't notice and continued skiing for the rest of the day.

Gabby Gingras was born with Hereditary Sensory Autonomic Neuropathy Type 5. This means she never developed nerve fibers that detect pain. Although this sounds like paradise, she gets injured a lot since she has little impulse to protect herself when she is severely injured. As a baby, she poked herself in the face so many times, she lost her vision in one eye. When Gingras was a toddler, she repeatedly bit her fingers until they were bloody.

She lost most of her teeth and broke her jaw by chewing on toys as child.

The disorder, Chromosome Deletion 6, caused Olivia Farnsworth to become incapable of feeling pain, fear, or hunger. She also needs very little sleep. In 2016, Farnsworth was hit by a car and dragged 30 meters. She got up without feeling pain and she only received minor injuries.

216. Panic Attack

A panic attack is a sudden period of intense fear. Although it usually lasts for minutes, it can go on for hours. Symptoms include hyperventilation, numbness, sweating, shaking, and palpitations. During a panic, the person tends to have paranoid thoughts like someone they love is going to die or they are losing control of their life.

A panic attack is usually caused by a repressed anxiety. Most people suffer their first panic attack over the most minor inconvenience e.g. missing a bus, can't find a phone charger, can't decide what to eat, etc. To an observer, having a panic attack over something minor seems like an overreaction. In reality, this "minor thing" is simply the straw that broke the camel's back.

Look at it this way – We learn early in life that we shouldn't get angry. As we get older, we try to hold back our anger as it is perceived as a negative emotion. Although anger can be detrimental, that doesn't change the fact that is a natural human reaction. If you hold back your anger or sadness or fear, it will not fade. Instead, it will get stronger. It will build and build and it will eventually snap like a coiled spring. This "snap" is what leads to a panic attack. The panic may be a once-off experience or it could take months or even years to correct.

Panic attacks are usually associated with people who are timid and weak. In reality, it's usually the opposite. Logical people usually develop a way to compartmentalize certain negative emotions like fear and rage. However, you can't trick your own brain. If you try to deny your emotions, they will seep out eventually.

Although panic attacks are usually caused by repressing emotion, they can develop from PTSD, depression, drug use, or a side effect to medication.

When someone is having a panic attack, people say things like, "Pull yourself together" or "It's all in your head." It's true that the panic is in your mind... but that doesn't mean the terror isn't real. Saying "snap out of it" to a person experiencing a panic attack will make the anxiety worse.

So, if that doesn't work, what is the best way to alleviate anxiety? Contrary to popular belief, breathing into a paper bag while having a panic attack isn't helpful. Taking in small, quick breaths are effective at alleviating stress.

Since panic is often initiated by holding back intense emotions, it's important for the person to be as honest about how they feel as possible. If you feel angry or stressed or scared, talk to your friends and family about it. If there is nobody around, verbalizing your insecurities is incredibly effective at conquering them. If you can't find what triggers the anxiety, therapy can help you out. There is medication to alleviate panic attacks but it wouldn't be advised unless the person has learned the origin of the problem. If they haven't, the medication will force the person to "sit" on the problem, making it worse.

Example – According to clinical psychologist, Dr. Ali Mattu, Iron Man 3 has a surprisingly accurate portrayal of a panic attack. Throughout the film, Tony Stark panics when people say certain words like "wormhole." These words are his trigger. He tries to ignore the anxiety which exasperates it and causes him to hyperventilate. However, Mattu points out that Stark's anxiety seems to just disappear in the film's conclusion, which isn't accurate.

In the show, This Is Us, Randall is a good example of a person who has panic attacks but is still able to function in society and has a successful career. Just because somebody suffers anxiety (or any mental disorder) doesn't mean their life is a trainwreck.

217. Paranoia
AKA Persecutory Delusions
Paranoia is classified as a psychosis that causes a person to form a generalized mistrust over others. Since the sufferer doesn't recognize they have a disorder, they often look out for clues to validate their suspicions that someone is out to get them. Trying to convince a person with paranoia to seek help tends to compound the problem more. A person with paranoia tends to suffer delusions of grandeur so they may see

themselves as a threat to an important figure or organization.

A person with paranoia may feel like others are watching them or trying to control them which may cause them to develop other disorders like agoraphobia or technophobia. They might become terrified that they will be poisoned and so, starve themselves.

Paranoia is very common with other personality disorders like narcissism, borderline personality disorder, and schizoid personality disorder.

Antidepressants and psychotherapy are effective treatments... if you can convince the person to agree to it.

218. Paraphilia

Paraphilia is a sexual arousal from atypical objects, people, or situations. Weirdly, a 2008 study committed on 200 heterosexual men showed that many paraphiliacs are left-handed and were the eldest in their family. Here are a few examples of paraphilia -

i)   Dacryphilia - Crying
ii) Spectrophile - Ghosts
iii) Gerontophilia - Elderly people
iv) Knismophilia - Tickling
v)   Agalmatophilia - Mannequins
vi) Acrotomophile - Amputees
vii) Dendrophilia - Trees
viii) Melissophilia - Animals
ix) Forniphilia - Turning a person into a piece of furniture
x)   Metrophilia - Poetry

219. Paris Syndrome

This disorder almost exclusively happens to elderly Japanese men. Many Japanese people dream of retiring to Paris after countless films have convinced them it is the most beautiful and romantic city in the world.

Sadly, building up the city for decades in their mind as the greatest place in the world causes Paris to be a complete letdown. Also, many Japanese people become confused and irritated as French customs are so different to Japanese ones. This condition is so common, Japan has a 24-hour emergency hotline for anyone suffering from Paris Syndrome.

220. Parkinson's Disease

AKA Shaking Palsy, Hypokinetic Rigid Syndrome, Idiopathic or Primary Parkinsonism, Paralysis Agitans

Parkinson's is a chronic disorder that attacks a person's central nervous system, which impairs motor functions. Over time, the person will develop symptoms such as shaking, slowed down movements, difficulty moving, rigidity, and difficulty collecting thoughts. It's likely that a person with Parkinson's will develop depression or anxiety. After being diagnosed, a Parkinson's suffer tends to live for ten years.

Doctors can detect Parkinson's in patients by looking for tremors, unstable posture, slow movement, and rigidity in wrist and elbow joints.

Although it's unknown what causes Parkinson's, it seems to occur more often with people with a history of head injuries. Weirdly, Parkinson's is most common in Guam.

Although it usually affects people in their 60s,

there are rare cases of people developing Parkinson's much younger. Back to the Future actor, Michael J. Fox, was diagnosed with Parkinson's when he was 30. Legendary boxer, Muhammad Ali, was diagnosed when he was only 42. The youngest record of Parkinson's sufferer was in a 12-year-old.

Men are twice as likely to develop Parkinson's.

Although there is treatment, it is expensive (about $2,500 per year,) and there is no cure.

221. Pathological Gambling
AKA Ludomania, Problem Gambling
The buzz that a person receives when they win money in gambling is identical to the buzz that a person experiences when they take cocaine. Gambling can be so addictive, players can bet their life savings on lotteries, casinos, or on races. Statistically, most gamblers are male alcoholics.

Pathological gamblers develop an addiction by falling into a trap known as the Gambler's Fallacy. This means that if a person keeps losing, he or she will convince themselves that they will win big soon to "balance it out." This makes a person develop unrealistic optimism that they are about to hit the jackpot even if they lose dozens of times.

There are many ways to treat pathological gambling including peer support, step-based programs, and anti-addiction drugs.

222. Pathological Generosity
This is a condition that forces a person to be absurdly nice. Although this condition sounds amazing, the affected person can become bankrupt since they

spend every penny buying gifts for others, even if they are total strangers.

223. Perfectionism
AKA Atelophobia
That's right. Being a perfectionist is classified as a disorder. More specifically, atelophobia is a fear of not being good enough. This can be advantageous as atelophobia can cause a person to strive to be successful in every aspect of their life. However, an atelophobic can become aggressive, depressed, or suicidal if they repeatedly fail.

224. Periodic Limb Movement Disorder
AKA Nocturnal Myoclonus
This disorder causes a person to move their limbs involuntarily while they sleep. People with PLMD also have difficulty staying awake during the day and sleeping at night. The disorder is usually connected to narcolepsy and Parkinson's. It normally affects people who work late shifts.

Anti-Parkinson medication, anticonvulsant, and benzodiazepine is effective at treating PLMD.

225. Persistent Genital Arousal Syndrome
PGAS causes a person to have a hundred orgasms a day. Cara Anaya may have the most intense version of this disorder as she can have 180 orgasms in two hours. That's an orgasm every 40 seconds. The orgasm can be triggered by a bump on the head or a phone vibrating. Many people who suffer this disorder find it unbearable as it's impossible for them to eat, sleep, or be a part of society. You can develop PGAS at any point

in your life and there is no cure.

226. Phantom Pain

The most common form of phantom pain is Phantom Limb Syndrome. PLS is when a person who is missing a limb has a sensation that their amputated limb is still attached.

Bizarrely, you can develop this disorder even if you have all your limbs! A person can have a pair of arms and legs but their brain is sending them signals as if there is a third arm or two extra legs. This is called Phantom Extra Limb Syndrome. This can be so extreme, some patients have accused hospitals of sowing an extra limb on them.

There's a similar disorder called Phantom Eye Syndrome where a person missing an eye can feel it moving around in their skull. If you lose an eye, that doesn't stop your brain sending signals to your optics. This can disorient your brain, causing you to see elongated objects. Sometimes, objects appear slanted or off centre.

There are many treatments including antidepressants, antiepileptics, electrical stimulation, hot and cold therapy, and massages.

227. Phantom Lover Syndrome

AKA Erotomania, Psychotic Erotic Transference.

This disorder causes a person to believe that a significant figure (often a celebrity,) is infatuated with them. This can happen even if the erotomaniac has never had any form of contact with the other person.

For unknown reasons, this disorder is much more common with women.

Examples – John Hinckley Jr was in love with the actress, Jodie Foster, and he tried to kill President Ronald Reagan, believing it would impress her.

The film, Fatal Attraction, revolves around Alex Forrest developing a murderous obsession with her lover, Dan Gallagher.

228. Phantom Odor
AKA Phantosmia, Olfactory Hallucination
This disorder causes a person to smell things that aren't there. It occurs when rogue neurons malfunction which causes them to send signals to the nose.

Phantom Odor definitely has one of the oddest remedies – cocaine. Seriously. Obviously, a phantosmiac can only have a very small amount, which is measured and monitored by a doctor. The drug, venlafaxine, is also effective.

229. Phantom Vibration Syndrome
AKA Fauxcellarm
PVS is disorder that causes a person to mistakenly believe their phone is vibrating. 68% of people regularly experience PVS.

230. Photographic Memory
AKA Hyperthymesia, Autobiographical Memory
This condition causes a person to be incapable of forgetting anything without effort. Jill Price has had this disorder since she was 14. Although this sounds like the jackpot of disorders, Price's photographic memory leaves her miserable. Not only does Price remember every insignificant thing in her life

perfectly, she also remembers every tragedy she has experienced with vivid clarity. Since the memories are bombarding Price every moment of every day, she find it very difficult to concentrate on anything else.

Example – One of the only examples of this disorder being depicted accurately in fiction is in the tv show, House MD. In Season 7 Episode 12, You Must Remember This, the doctors deal with a waitress who suffers hyperthymesia.

231. Pica

Pica causes a person to develop an urge to eat anything that isn't food. This disorder usually begins from a young age and can last a lifetime. It's especially common with pregnant women and autistic people.

Types of pica include –

i) Pagophagia - Ice
ii) Trichophagia – Hair (This disorder is also known as Rapunzel Syndrome.)
iii) Xylophagia – Paper and wood
iv) Mucophagia - Mucus
v) Lithophagia – Stones
vi) Geophagia – Dirt
vii) Acuphagia – Razor blades
viii) Hematophagia - Blood
ix) Hypalophagia - Glass
x) Autophagia – One's own body

Pica is usually caused by a mineral deficiency. Treatment depends on what a pica sufferer eats. If they are drinking blood, it may be because they suffer an iron deficiency. If that is the case, the person should receive iron supplements.

Example – Michael Lotito was known as Mr. Eat-All

since he ate bicycles, shopping cats, televisions, razor blades, plates, etc. He ate a Cessna 150 plane over a period of two years. Lotito suffered no ill effects from eating these "foods." Weirdly, normal foods like eggs and bananas made him sick.

232.   Pneumonia
Pneumonia is an infection of the lungs. As pneumonia takes hold, the lungs fill with fluid, causing the person to develop a dry cough, fever, breathlessness, and chest pain.

Pneumonia is usually caused from another illness like asthma, the flu, or chronic obstructive pulmonary disease.

Pneumonia can develop into the lethal disease, sepsis. 450 million people develop pneumonia every year, killing four million. It is the leading cause of death worldwide among children under the age of five years old. The physician, William Osler, said pneumonia was "the captain of the men of death." Since it is usually a secondary illness for the elderly, pneumonia is also known as "the old man's friend."

Although there are vaccines to prevent certain types of pneumonia, the illness can be avoided with regular handwashing. If pneumonia sets in, it should be fine if treated early. Garlic and lemon water can alleviate the disorder.

233. Polio
AKA Infantile Paralysis, Poliomyelitis
This disease weakens muscles, causing the person to have impaired mobility and strength. Although polio usually effects the legs, the disease can affect the

hands and neck. Polio was usually transmitted from contaminated milk. Originally, scientists believed ice-cream was the cause of polio.

Once polio sets in, there is no cure. Developing polio nowadays is astonishingly rare thanks to the Salk vaccine. The vaccine has to be taken several times to be effective. In the 1950s, some people accused the polio vaccine to be a Communist plot to socialize medicine. As of 2018, polio has only been recorded in Afghanistan, Nigeria, and Pakistan.

Jonak Salk didn't patent the vaccine as he believed all medicine should be free. His generosity saved millions of lives.

234.   Polydactyly
This disorder causes a person to have extra fingers or toes. Devendra Suthar has the world record for polydactyly since he has 14 fingers (including four thumbs) and 14 toes.

Oligodactyly is the opposite disorder where the person has fewer digits.
Example – In the animated series, Gravity Falls, Ford has six fingers on one hand. In the novel, The Silence of the Lambs, Hannibal Lecter has an extra finger.

235.   Polycystic Ovary Syndrome
Although the name of this disorder implies the affected woman has cysts on her ovaries, that is not always the case. PCOS is when a woman has an overabundance of androgens (male hormones.) This can lead to excessive body hair, obesity, diabetes, irregular or no periods, pelvic pain, acne, and difficult

pregnancies. 40% of PCOS sufferers will develop diabetes before they are 40.

Although PCOS is one of the most common endocrine disorders, it receives less than 0.1% of funding from the National Institutes of Health.

PCOS affects 8% of people in the developed world. The condition is usually caused by genetics. Although there is no cure, a doctor can recommend medication to regulate the androgen levels. If a woman does not receive regular medication for PCOS, she is more susceptible to ovarian cancer.

236.   Postnatal Depression
AKA Postpartum Depression, Baby Blues
PPD is a mood disorder that 15% of women suffer after giving birth. The symptoms include anxiety, panic attacks, de-energization, spontaneous crying, irregular eating and sleeping pattern. Women with PPD usually experience an intense dread, similar to what bipolar sufferers endure. Postnatal usually lasts for a few weeks but it can last months.

Despite being a very common disorder, no one is sure what causes postnatal. Counselling and medication are notable treatments for PDD.

237.   Post-Traumatic Stress Disorder
AKA Shell Shock
PTSD is when a person is traumatized by flashbacks. These flashbacks are often connected to serious injury, sexual violence, or death. Although PTSD has been documented in ancient Mesopotamia, it didn't become common knowledge until World War I. During battle, soldiers developed tremors, fatigue, and

disorientation. At the time, it was believed that soldiers' behavior was due to shock after being exposed to detonated explosives or "shells." When shell-shock was recognized as a disorder, it was believed to be a physical condition, not a mental one. After this condition was understood to psychologically affect the patient and it was not only applicable to soldiers, it became known as Post-Traumatic Stress Syndrome.

Left-handed people are far more likely to suffer from PTSD.

238. Prader-Willi Syndrome
AKA Labhart-Willi Syndrome, Fanconi Syndrome
PWS is a genetic disorder that causes slow development, weak muscles, excessive sleep, and incredible hunger pangs, which causes the affected person to put on weight from a very early age. This often leads to type 2 diabetes. 74% of cases occur when part of the father's chromosome 15 is deleted. Sadly, there is no cure.
Example – Model, Katie Price's, son, Harvey suffers PWS as well as blindness and autism.

239. Pronoia
Pronoia is the opposite of paranoia. A person with pronoia believes that people are conspiring to help them. "Pronoia" is Greek for "care" or "forethought." Some psychologists associate pronoia with delusions of grandeur since a person with pronoia may assume that everyone likes them when there is clear evidence of the opposite.

240. Proprioception

This condition causes the sufferer to lose the ability to know where their own body parts are. When you wake up, you are probably still half-asleep and you can't think straight until you get your morning coffee. Even if you feel tired, you can get dressed, have a shower, eat breakfast, brush your teeth, etc. You don't have to consciously make these movements because it happens instinctively.

However, a person with proprioception cannot do this. A patient with this illness has to be conscious of every movement they are making, which makes bending, turning, standing, and walking a constant struggle.

"Proprioception" is Latin for "to grasp one's own."

241. Psoriasis

Psoriasis is an autoimmune disease that causes a person to develop a patchy rash on their skin. It is unknown what causes psoriasis. The rash usually affects the elbows, lower back, and knees. There are five types of psoriasis –

i) Eruptive psoriasis is caused by bacteria.

ii) Inverse psoriasis forms in the folds e.g. ears, groin, and lips.

iii) Pustular psoriasis has a blistery texture that effects the hands and feet.

iv) Erythroderma effects the entire body.

v) Plaque psoriasis has a red or purple rash that can take on a smooth or scaly texture. 90% of psoriasis sufferers have this version.

Approximately 2.5% of people suffer from this disease. The name is derived from the Greek word "psora" which means "itch."

Coconut oil is effective against psoriasis. As hard as it is to believe, chemotherapy is effective against chronic psoriasis. Weirdly, the oldest recorded remedy for psoriasis was urine.

It didn't work. Ever.

242. Psychosomatica
AKA Somantic Symptom Disorder

A psychosomatic illness is a physical ailment that manifests mentally. Although this is often mixed up with hypochondria, psychosomatica is very real. A hypochondriac believes they are ill when they are not. Psychosomatica is when a healthy person is so stressed, they become ill. Under intense stress, a person can develop a headache, a cold, fatigue, a rash, or eczema.

The most extreme version of psychosomatica is Conversion Disorder. When a person is under the highest level of stress, they can go mute, blind, or paralyzed, even though these symptoms can't be traced to a medical cause.

Although cognitive behavioral therapy and anti-depressants are effective treatments, it's necessary to acknowledge what is causing the stress and finding a way to minimize it.

243.   Psychopathy

A psychopath is incapable of feeling empathy. Society usually sees psychopaths as people who are incapable of feeling emotion. This is utterly wrong. A psychopath is prone to anger, fear, joy, and sadness like anyone else. They just can't feel compassion for other people's emotions. Psychopathy was the first condition to be referred to as a "mental illness."

There are four types of psychopath -

i) Disaffiliated psychos have no connection to anyone on an emotional level. These are the psychopaths that are usually depicted in films and television series.

ii) Hostile psychos are easy to anger and prone to violence.

iii) Disempathetic psychos can function socially in certain ways but are still prone to erratic behavior. Richard "Ice Man" Kuklinski appeared to be a loving husband who cared for his children despite being a hitman who killed dozens of people. Everyone who knew Kuklinski said he seemed like the last person they would suspect of committing murder.

iv) Cheated psychos feel a deep-rooted hatred against a person, a gender, a religion, society, etc. and they will exact their revenge on anyone they seem fit.

Although psychopaths like Hannibal Lecter in The Silence of the Lambs are often depicted as criminal masterminds, the majority of them are of below-average intelligence.

Most psychopaths aren't violent or dangerous. It is the very few psychopaths on the extreme end of the spectrum that makes the word associated with "serial killer."

It is also worth mentioning that psychiatrists never officially diagnose patients as "psychopaths" because the term is too broad. The closest diagnosis is Antisocial Personality Disorder.

244. Pyromania

Pyromania is an impulse control disorder that compels a person to set fires. However, a pyromaniac is different to an arsonist. Arsonists believe the act or arson will have some personal or financial gain. A pyromaniac sets fire for a buzz or to feel in control. They rarely set fires to cause harm to others.

The root of pyromania usually lies in the person having a lack of control over at least one aspect of their life. Cognitive behavioral therapy is effective at treating pyromania.

245. Rabies

Rabies is a viral disease which causes an inflammation in the brain. The word "rabies" is derived from the Latin for "rage," since the disease causes the infected to act wild.

Although the disease occurs in people, chickens, bats, and foxes, rabies usually occurs in dogs. Rabies causes animals to become feral, meaning that they will try to attack anything that comes near them. Because of this, infected dogs have to be put down.

The concept of rabies became common knowledge from the film, Old Yeller. In the film, Old Yeller becomes ferocious after developing rabies and so, is put to sleep. In reality, animals with rabies don't act wild. Normally, they appear drunk and uncoordinated. Because the animal usually looks injured, people try to "rescue" it, allowing themselves to be infected.

A human can develop rabies if they are bitten or are exposed to the fluid of an infected animal. A person with rabies can develop a fever, hallucinations, spasms, and paralysis. An infected person may develop hydrophobia, which makes them absolutely terrified of water. They will do everything in their power to avoid drinking liquid, even if they are dehydrated. Even the thought of water can caused an infected person to have intense throat spasms.

Luckily, human rabies is astoundingly rare. In 2010, only two people in the US were diagnosed with rabies.

If a person is infected with rabies, they should be fine if they receive medical treatment immediately. Once the symptoms become apparent, the result is nearly always death.

Although rabies is usually spread to humans through dog saliva, only 5% of rabies cases in the US are caused by canines. The most likely animal to spread rabies to humans in the Americas is the bat.

246. Rapid Eye Movement Sleep Behavior Disorder

This disorder is the opposite of sleep paralysis. When a person with this illness falls asleep, the brain doesn't send the signal to paralyze the muscles. This causes the person to physically act out their dreams. They will also speak even though they are asleep. If the person is having a nightmare, they can cause severe injuries to themselves or anyone nearby.

Brian Thomas had a dream that he was attacking a group of burglars in his house. When Thomas awoke, he learned that he had strangled his wife to death. Since Thomas was known for this condition, he was acquitted of the crime.

People with this disorder usually suffer neurodegenerative problems. It's also not uncommon with the elderly.

247.   Recollective Confabulation
AKA Disordered Recognition Memory
RC is when the brain registers new information as familiar information. A person with RC may feel like they recognize someone they just met for the first time. A RC sufferer might enter a new country but feel like they have been there before. RC is like a very intense version of déjà vu.

248. Reduplicative Paramnesia
This delusion makes the patient believe that a person, object, or location has been duplicated. If a person develops this disorder while in their house, they may feel like they are not really in their home but an identical building. A paramnesiac's delusion can be so strong, they may dismiss any evidence such as observing that the garden, front lawn, neighboring houses, and streets are identical.

This condition tends to happen to someone who received an injury to the right cerebral hemisphere and both frontal lobes.

249. Renfield Syndrome
AKA Clinical Vampirism
Renfield Syndrome causes the person to be obsessed with drinking blood. It's named after the character, Renfield, from Bram Stoker's novel, Dracula. In the book, Renfield had a habit of eating rodents. RS is very rare and is nearly always a side effect to another

disorder like PTSD or DID.

250. Resignation Syndrome

This is a disorder that causes the sufferer to go into a coma-like state when they learn that a family member is to be deported. Bizarrely, this only happens to one group of people in the world – Yugoslav and Russian refugee children who reside in Sweden. Although this might sound like the children are sulking or protesting, they didn't decide to do this. When family members in Sweden were being deported during the 1990s, children across the country suddenly locked up and refused to leave their beds.

And this didn't go on for a few days. Some children were in this condition for over two years. While in this state, the children didn't eat, speak, move, or react to stimuli. When they eventually left their beds, it took them months to recover physically and mentally.

One doctor described the sufferers as "Snow White" who "just fall away from the world."

251. Restless Legs Syndrome
AKA Wittmaack-Ekbom

This disorder is characterized by itching, cramping, and tingling while a person is lying down.

Although comedian, Ricky Gervais, makes fun of his friend, Karl Pilkington, for suffering this "made-up-illness," RLS effects 7% of people. If a woman develops RLS when she is pregnant, she should consult a doctor as soon as she can as it can develop into a permanent form of the disorder. RLS can be curbed with exercised or treated with medication.

252. Retired Husband Syndrome
AKA One's Husband Being at Home Stress Syndrome
RHS effects up to 60% of women in Japan. In Japan, women become so used to their husbands at work for decades, they find it jarring when he retires because his presence in the house messes up his wife's routine. This illness has been known to cause physical ailments like stomach ulcers and rashes.

253. Riddoch's Phenomenon
This condition makes a person unable to see objects unless they are moving. 29-year-old, Milena Channing, was told by her doctor that she went blind after she suffered a stroke. While she was drinking her coffee one day, Channing could see the steam from her drink. However, she couldn't see the cup that her coffee was in. Realizing that she couldn't see objects unless they appeared to be moving, Channing decided to get a rocking chair. As she rocks in her chair, Channing gains a semblance of her sight back.

254. ROHHAD
In 2017, Jake Vella was only eight-years-old when he started competing in triathlons. If that sounds impressive, Jake accomplishes this despite the fact that he is obese and has a tumor on his back. So, if he is so overweight, how can he perform such a physical feat?

Because Vella has to or he will die. Vella suffers from ROHHAD (Rapid-Onset Obesity with Hypothalamic Dysregulation, Hypoventilation and Autonomic Dysregulation.) Less than a hundred people have been diagnosed with this disorder. Sadly, no sufferer has survived beyond their 20s.

255. Rubber Skin
AKA Ehlers-Danlos Syndrome, Sack-Barabas Syndrome

EDS affects the skin, joints, and blood vessels of a person's body. However, EDS is best-known as the disorder that causes a person to develop overly elastic skin. EDS sufferer, Garry Turner, has skin that is so loose, he can stretch it out like a rubber band. As a party trick, he covers his entire mouth with the skin of his neck.

A person with EDS suffers permanent joint pain, random dislocations, and nausea. Worse still, sufferers have a high resistance to anesthesia. Some of them have an immunity to it. Sadly, there is no cure for this syndrome.

256. Rubella
AKA German Measles, Three-Day Measles

Rubella is a highly contagious viral infection that causes a rash to spread throughout the body for several days. Although rubella sounds scary because it can cause birth defects if it infects a pregnant woman, it is one of the easiest diseases to vaccinate against. The MMR (Measles, mumps, rubella) vaccine is given to babies when they are six months old. The child usually receives a booster just before they start school.

257. Sadness Immunity
After suffering a stroke, Malcolm Myatt became incapable of feeling sad. There are only a handful of

cases of this occurring but it has only been reported with people who have suffered a stroke.

258. SARS

Severe Acute Respiratory Syndrome is a viral disease that created a scare from November 2002-July 2003 when 8,098 people in southern China were infected. SARS eventually spread to other countries, which led to the death of 774 people. There have been no reports of anyone being diagnosed with SARS since 2004.

The disease causes the infected to develop a fever, sore throat, and muscle pain. Sometimes, the person develops pneumonia, which is what most SARS sufferers died from.

In 2017, scientists learned that the disease originated from Horseshoe bats.

259. Scarlet Fever
AKA Scarlatina

This disease is triggered by a germ called A streptoccus. Certain people are immune to this bacterium. Two kids in the same family can be exposed to this germ but only one of them develops scarlet fever. It causes the sufferer to develop a sore throat, fever, headache, a white coating on the tongue, and swollen lymph nodes. The most distinguishable feature is a red bumpy rash where the person's skin feels as coarse as sandpaper. Although the fever usually lasts a week, the person's skin may peel for up to a month.

This condition can be passed on if a person has direct contact with saliva, mucus, or the skin of an infected person. It usually affects children between the ages of 2-10.

In the early 20th century, people with scarlet fever tended to develop arthritis, abscesses, and pneumonia. Although this was a potentially lethal disorder during Victorian times, modern medicine has caused scarlet fever to be very treatable. However, any child should stay at home from school for at least 24 hours after beginning their antibiotic treatment.

260.   Schizoid Personality Disorder

A person with SPD tends to have little to no interest in social relationships. They act aloof, secretive, detached, and seem to be immune to criticism. Despite this disorder's name, SPD is not connected to schizophrenia.

Example - Kramer from the tv series, Seinfeld.

261.   Schizophrenia

This mental disorder causes the affected person to hear sounds or voices that don't exist. A schizophrenic may hear a voice speaking to them even if they are completely alone. The voice may sound normal or otherworldly.

A sleeping schizophrenic can be woken up by one of these voices. Many religious people believe these voices are from angels, Christ, God, the Devil, or a ghost.

This disorder can develop in someone if they used cannabis excessively during adolescence. Schizophrenia can also be caused due to childhood trauma or genetics. Weirdly, many schizophrenics can tickle themselves.

Contrary to popular belief, schizophrenia is not connected to dissociative identity disorder or physical hallucinations.

The Schizophrenia Research Journal showed that 50.6% of schizophrenic adults had a cat as a child. Men are more likely to suffer schizophrenia and suffer more side effects than women.

Once schizophrenia develops, it's there for good. Over the years, the disorder can make a person depressed, aggressive, or suicidal.

Interestingly, there is no record of a blind person developing schizophrenia.

Schizophrenics tend to take antipsychotic medication to suppress the voices but meditation or counselling are also effective.

Example – Russell Crowe plays real-life Nobel Prize winner, John Nash, in the 2001 film, A Beautiful Mind. Although John Nash suffered auditory hallucinations, the film depicts him having visual hallucinations which is erroneous. However, this is one of the very few films that depicts a schizophrenic as non-violent, which is accurate.

262. Sciatica

Sciatica is an excruciating pain going down the leg from the lower back. It normally develops after intense exercise or heavy lifting. It usually only effects one side of the body. 90% of the time, it is triggered by a spinal disc pressing into the lumbar or sacral nerves. Although the condition has been known for millennia, physicians only learned in the early 20th century that sciatica was caused by herniated discs.

Men are three times more likely to develop sciatica

than women.

The pain is so bad, the sufferer will instinctively want to rest. This is one of the worst things you can do. Although it seems counter-intuitive, it's vital for a sciatica-sufferer to stay active by performing basic tasks. If they rest, the discs will go into a hibernation-like state, which can take much longer to heal.

Chamomile tea helps alleviate sciatica. In Ancient Greek times, the boiled milk of a donkey was used against sciatica. Personally, I would stick with the chamomile tea.

263.   Scoliosis

Scoliosis is a disorder that causes a sideways curve in the back. Although scoliosis affects 3% of people in the developed world, it's ten times more likely in women. It can be caused by an injury, cerebral palsy, spinal tumors, or Marfan syndrome. Scoliosis is usually seen as an inconvenience rather than a life-or-death disorder. Surgery is only considered if the scoliosis is severe.

The disorder has first diagnosed by Ancient Greek physician, Hippocrates. He named the disorder "Scoliosis" which means "bending."

Scoliois occurs in dogs, cats, horses, and fish. Weirdly, gorillas and chimps don't suffer scoliosis despite their genetic similarities to humans.

Examples – Usain Bolt, Sarah Michelle Gellar, Elizabeth Taylor, Kurt Cobain, and Liza Minelli had scoliosis. Based on sculptures and paintings of Alexander the Great, some historians believe that he suffered from scoliosis.

264.   Scurvy

AKA Moeller's Disease, Cheadle's Disease, Barlow's Disease, Scorbutus, Hypoascobemia

If a person has a lack of Vitamin C in their body, they will develop scurvy, which causes the gums and tongue to bleed. If the disorder is not resolved, the person will develop a lethal infection.

Although this disorder was incredibly common in Christopher Columbus' time, scurvy is rare nowadays since food is so accessible. However, scurvy is not uncommon in newborn babies or the elderly.

If scurvy sets in, it is vital to consume as much Vitamin C as possible before the symptoms exasperate. Although strawberries, oranges, lemons, and papayas have over 50g of Vitamin C, blackcurrants have 181gm, guava has 228g, and the urtica plant has a staggering 333g of Vitamin C.

265. Seasonal Affective Disorder

This mood disorder causes the person to become depressed or anxious during winter. Although experts were originally skeptical about SAD, it is now classified as a disorder. It's most common in people who are from a warm country who move to a country that is regularly cold or raining. Because the person is not exposed to the same amount of sunlight, they are not receiving the level of Vitamin D they are accustomed to. This lowers the person's serotonin levels, causing them to feel fatigued, agitated, or anxious. Since bipolar sufferers have a serotonin deficiency, it's not uncommon for them to develop SAD.

If SAD is drastically affecting a person's life, they

should receive medication, light therapy, or melatonin supplements.

266. The Self-Diagnosing Tumor

In 1984, a woman in London was reading when she suddenly heard a voice saying, "Please don't be afraid. I know it must be shocking for you to hear me speaking to you like this, but this is the easiest way I could think of. My friend and I used to work at the Great Ormond Street Hospital for Children and would like to help you."

Although the woman assumed she was daydreaming, the voice persisted and so, she went to the doctor and explained her situation to Dr. Azunoye. The woman told the doctor that this voice said she had a brain tumor and an inflammation on her brain stem. Not only did a brain scan confirm this, but the woman had no other symptoms that a doctor could pick up on. After she had the operation, the woman heard the voice say, "We are pleased to have helped you." That was the last time she heard the voice.

Azuonye suggested that the growing tumor in her brain caused enough residual sensation for the patient to be aware that something was wrong, which manifested as a mysterious voice. The fact that the voice disappeared after the operation suggests that the disorder was linked to a neurological disorder.

Example - Mark Ruffalo, who is best known for playing the Hulk in the Marvel films, had a dream that he had a brain tumor. Paranoid, he went to the doctor to get a scan, which confirmed his suspicions. Ruffalo had it removed and he eventually recovered.

267. Senile Squalor Syndrome
AKA Diogenes Syndrome
This disorder causes a person to live in squalor and makes no effort to socialize or look after themselves. Hoarding is also a common trait for people with SSS. The person may have a dozen pets or more but make little to no effort to care for them.

268.   Sepsis
Sepsis is an infection that causes germs to multiply in the body, damaging tissue and organs. Sepsis can develop from virtually any infection and it can affect any person of any age. According to the National Institute of General Medical Sciences, sepsis is usually caused by E. coli, staph, strep, fungi and viruses.

Although the disorder has been referred to as "septicemia" and "blood poisoning" since Hippocrates' time, these terms are considered outdated.

Sepsis is difficult to identify since many of its early symptoms don't seem serious. At first, a person with sepsis will develop clammy skin, shortness of breath, dizziness, and a sore throat. With little warning, it can cause inflammation, which causes blood clots and leaky vessels, which prevents oxygen from reaching vital organs. If an organ fails, the body goes into septic shock

Sepsis is treated with antibiotics and intravenous fluids. It is vital to get treated for sepsis as soon as possible as it is the leading cause of death in ICU.

Only 0.25% of people develop this disorder per year. Although that doesn't sound like a lot, that's 750,000 Americans developing sepsis annually. Worse still, sepsis kills 28-50% of people it infects. On top of that, sepsis has been increasing in recent years.

Receiving a vaccination from the flu and pneumonia are critical to stopping sepsis from developing but it is not a guarantee.

269. Sexsomnia

Sexsomnia causes a person to try to have sex while he or she is asleep. In 2016, a Swedish man was cleared of rape charges after it was revealed that he suffered from sexsomnia.

Anticonvulsant therapy and medication is recommended for treating sexsomnia.

270. Shared Psychosis

AKA Folie a Deux, which translates into "madness of two."

This is when more than one person share a delusion. It tends to happen to two people who are married, related, or live together. Depending on how many people share the delusion, it can be referred to as Family Madness or Madness of Several.

Example - The first record of a Shared Psychosis was by a couple called Margaret and Michael during the 19th century. They shared the delusion that people were breaking into their house, spreading dust everywhere, and wearing down the couple's shoes.

271. Shingles

AKA Herpes Zoster

This viral disease develops a blistery skin rash, usually as a large stripe on the side of the face or body. It is often accompanied with postherpetic neuralgia, which can leave the sufferer with nerve pain for months or years.

Shingles usually develops from a reactivation from chickenpox. If you never had chickenpox as a child and develop it as an adult, you can get shingles.

The shingles vaccine reduces the risk of the disease by 50-90%.

272. Shy Bladder Syndrome
AKA Paruresis

This is an anxiety disorder that prevents a man to urinate in public. Approximately 7% of males suffer this condition. It can develop if the person had a bad experience while being potty-trained or a humiliating experience while they were in the bathroom at some point in their life.

273. Skin Hunger

This is an anxiety that one develops due to receiving little to no physical contact with other people.

274. Sleep Apnea

Sleep Apnea causes a person to stop breathing while they sleep. This doesn't just happen once or twice. A person with apnea can stop breathing dozens of times during the night. There are two forms of apnea – Obstructive Sleep Apnea and Central Sleep Apnea.

OSA is when the throat muscles relax, which closes the airways.

CSA is when the brain stops sending messages to the breathing muscles. This disorder can be caused from high blood pressure or obesity.

This potentially lethal condition affects 18 million Americans. Some people with this condition have to wear an air pressure mask to maintain their breathing.

275. Sleep-Eating

Sleep-Eating compels people to binge-eat in the middle of the night. Although everyone has had a midnight snack at some point, people who sleep-eat may not even remember getting up in the middle of the night.

This can be a lethal disorder since the sufferer can eat toxic material while sleeping. There have been reports of SRED sufferers chomping on cat food, salt sandwiches, and cigarettes. For some reason, SRED usually affects middle-aged women.

276. Sleep Paralysis

When a person is asleep, certain muscles are paralyzed to stop them from hurting themselves by acting out their dreams. However, a person has wake up and find themselves unable to move. This is known as sleep paralysis. This can last for minutes or hours and the person won't be able to speak until it wears off.

Sleep paralysis hallucinations can be visual or auditory. Visual hallucinations usually take the form of a shadowy entity or a monster. Auditory hallucinations are normally screeching or howling. These hallucinations can psychologically affect the person after they awake for hours or days.

Persians believed sleep paralysis was caused by a demon called Bakhtak. This entity sits on his victim while they sleep and turns their dream into a nightmare. If the person awakens from the nightmare, Bakhtak will paralyze them so he can get away.

This condition usually happens to people sleeping on their back. To limit the risk of sleep paralysis, one should sleep on their side.

Antidepressants, serotonin inhibitors, and cognitive behavior therapy are good treatments for sleep paralysis.

277. Sleep State Misperception

This disorder causes a person to believe they are awake when they are asleep. They believe they have been awake for days or weeks, oblivious that they have been sleeping.

Example – Trevor Reznik in the film, The Machinist.

278. Sleeping Beauty Syndrome

AKA Rip Van Winkle Disorder, Kleine-Levin Syndrome

This causes people to sleep for 23 hours. However, some people with SBS can sleep for days. The person may suffer SBS for weeks or months. When awake, they are usually extremely hungry (probably because they didn't eat for a day.)

Nobody knows what causes it although scientists say many sufferers developed an infection just before the SBS kicked in. Sleeping Beauty Syndrome usually happens to teenage boys.

Although there is no permanent cure, lithium works as a temporary treatment.

279. Sleepless Disorder

Some people have a mutation in the gene SNP rs121912616 that allows them to sleep two hours per night with no side effects.

280. Sleepwalking

AKA Somnambulism, Noctambulism

Sleepwalking is the act of moving around while one is asleep. Sleepwalkers usually have no memory of the activities they perform during this time. It is most common in people between the ages of 4-12. According to a 1997 study published by The Journal of Neurology, sleepwalking occurred in 3.9% of men and 3.1% of women.

Interestingly, the word "zombie" is derived from the word "somnambulism," as a sleepwalker and a zombie move in a slow mindless manner.

If either of your parents sleepwalk, it's more likely that you or your siblings will do so.

Sleepwalking doesn't happen randomly. It is triggered by something like medication, alcohol, a night terror, or a trauma.

A somnambulist's eyes may or may not be open while they are sleepwalking. It's very difficult to tell if someone is sleepwalking as they may be able to respond to another person. They can perform surprisingly complicated acts including going to the toilet, making a meal, holding a conversation, playing a musical instrument, or driving. There have been some cases where a person committed murder while sleepwalking.

Some sleepwalkers dream while others don't.

There is a myth that you are not supposed to wake a sleepwalker. This is ridiculous. If you saw a sleepwalker about to walk on something dangerous, you should obviously wake them. They will be disorientated but that's better than being injured. This is a rare occurrence as most somnambulists have no bad experiences during their sleepwalk. Just to be on the safe side, the most logical thing to do if you see a sleepwalker is to guide them back to their bed.

Sleepwalkers will usually feel extremely tired the following morning since their sleep cycle was interrupted.

281. Smallpox
AKA Variola, The Red Plague
Smallpox is a highly contagious disease that causes the infected to develop a fever and mouth sores. As the disease worsens, the person develops fluid-filled blisters on their skin, which eventually scabs over. Smallpox had a 30% fatality rate. Anyone who survived was usually rendered blind.

During the 20th century, it killed about 400 million people. As recent as 1967, there were 15 million reported deaths from the disease.

Despite how devastating smallpox is, it's mainly known for one thing – It doesn't exist anymore. Smallpox is one of two diseases to be completely eradicated thanks to vaccines. (In case you were wondering, the other disease is Cattle Plague.) Smallpox was eradicated in 1977. For research purposes, a batch of smallpox is sealed away in the CDC in Atlanta.

The last person to contract smallpox, Ali Maalin, dedicated his life to eradicating polio.

282. Smile Mask Syndrome
Although it's human nature to smile when you are happy, this can be a problem if you have a job where you are forced to smile all the time. In 1983, a

Japanese psychologist noticed a disorder called Smile Mask Syndrome which affects people who have to smile for work e.g. air hostess, waitress, entertainer, etc. They became so deadened to smiling, they couldn't feel happiness from it and developed depression and physical illnesses.

283. Somatic Symptom Disorder
Somatic Symptom Disorder is when a person develops a physical illness for no apparent reason. A person can develop a bruise, which looks like it was caused by injury but it may have manifested itself overnight. A patient may develop a full-body rash but a medical exam will state that the person is completely healthy. As a result, it's likely the doctor will misdiagnose the patient or accuse them of hypochondria.

284. Somatoparaphrenia
This is a disorder that convinces the sufferer that one or several limbs don't belong to them. Some somatoparphrenians believe one of their limbs is an alien implant. This neurological disorder is so severe, it's almost impossible to convince the sufferer that what they believe is untrue. In fact, some sufferers were so convinced their limb wasn't their own, they had it amputated.

285. Species Dysphoria
This disorder causes a person to believe they were born as the wrong species.

286. Spina Bifida
Spina bifida is a birth defect that stops the backbone

from sealing around the spinal cord. "Spina bifida" literally means "split spine." It usually occurs in the lower back. There are three types of SB.

    i) Occulta is a swelling or dark spot on a gap of the spine. 15% of SB patients have occulta.

    ii)   Meningocele causes a sac or fluid at the spine gap.

    iii)   Myelomeningocele or Open SB is the most severe form of this birth defect. Open SB causes impaired mobility, incontinence, and fluid in the brain. It's also likely the person will develop an allergy to latex.

Although it is uncertain what causes spina bifida, researchers say that women with a low amount of folic acid and Vitamin B tend to have a child with the condition. Contrary to what many people believe, the condition doesn't impair the person's learning abilities.

According to Newborn Surgery, Caucasians are much more likely to develop spina bifida than black people.

Although there are treatments for the condition, there is no cure. Depending on the symptoms, the person will need to be rehabilitated with a neurologist, physical therapist, urologist, or orthotist.

287. Sprain
AKA Torn Ligament
A sprain is caused when the joints in a ligament are damaged when it's forced to move beyond its functional range of motion. This usually occurs when the joint is struck with great force. Despite what some

people believe, you can't sprain a muscle or a tendon. Although sprains are considered a minor problem since it usually heals after a few days, it can take years depending on the severity of the injury. Some sprains can only be fixed with surgery.

Although the ankle is the most common ligament to sprain, you can sprain the knee or wrist.

It is important not to put any pressure on the sprained ligament. Walking should be kept to a minimum. Ice should be placed on the injury as soon as possible but for no longer than 15 minutes at a time. This should be performed three to four times daily. Keeping the sprain elevated will slow down the swelling. If you have to move with a sprain, it is advised to dress the injury with bandages or ace-wraps. When wrapping, apply more pressure at the far end of the injury to avoid cutting off circulation of the limb. Depending on the severity of the sprain, you should use crutches or a knee scooter.

288. Stereo Blindness
Stereo blindness prevents people from seeing in 3D. This means that everything they see appears flat. Individuals with one eye always have this disorder.

289. Stockholm Syndrome
AKA Norrmalmstorg Syndrome
This psychological condition causes a hostage to befriend their captor. The term was originally used in 1973 when four hostages were taken during a bank robbery in Stockholm, Sweden. When the criminals were arrested, the hostages defended the robbers' actions and refused to testify against them in court.

According to two studies performed by the FBI, 8% of hostages develop Stockholm Syndrome during a hostage situation.

There are four factors to Stockholm Syndrome -

i)   The captor and the hostage don't know each other

ii) The hostage develops positive feelings towards the captor

iii) The hostage refuses to co-operate with authorities

iv) The hostage understands the captor's motivation and may believe it is justified

Example – Belle in Beauty and the Beast

290.   Stone Man Syndrome

AKA Fibrodysplasia Ossificans Progressiva

This is an extremely rare disease that attacks the connective tissue. A mutation of the body's repair mechanism causes fibrous tissue (including muscle, tendon, and ligament) to be ossified spontaneously or when damaged. This means that these tissues will turn into bone. In many cases, injuries can cause joints to become permanently frozen in place.

The affected person has greater difficulty moving, talking, and eating every passing day until they resemble a living statue. Not only is there no cure but attempting to remove these bones simply magnifies the speed that the cartilage grows at. This disease only affects 0.00005% of the world's population.

291. Stroke

AKA Apoplexy, Brain Attack, Cerebrovascular Accident

When blood flow to the brain is blocked, it results in

cell death. This is known as a stroke. There are two types –

i) An ischemic stroke is caused by a lack of blood flow. 70% of strokes are ischemic.

ii) A hemorrhagic stroke is caused by bleeding.

Symptoms of a stroke include impaired movement or feeling on one side of the body, loss of vision in one eye, irritability, slurred speech, and difficulty understanding basic concepts. These symptoms may last a few months or they can be permanent.

It is more likely you will have a stroke if you have high blood pressure, high cholesterol, diabetes, or if you smoke or drink alcohol regularly. Strokes are the fifth most common cause of death in the US.

Although men are more inclined to suffer a stroke, it's more fatal to women. In fact, a stroke is more likely to kill a woman than breast cancer.

Bizarrely, it's possible to develop a stroke while getting one's haired washed at a salon. That's not a joke. If the head is tilted back too long, it tears the neck artery. This has happened so many times, it's nicknamed Beauty Parlor Syndrome.

In terms of treatment, aspirin is effective against an ischemic stroke. If the stroke is severe, surgery may be needed to remove the clot.

For a hemorrhagic stroke, the patient's blood pressure, blood sugar levels, and oxygenation must be regulated. Depending on the position of the clot, surgery may be necessary.

Stroke victims will need a lot of physical rehabilitation through exercise and electrical stimulation on certain muscles. Psychological rehabilitation is also vital since many stroke victims become prone to depression or irritability. The stroke victim will need cognitive therapy if he or she has suffered memory impairments. According to one American study performed on a group of stroke victims over the age of 65 –

50% were paralyzed on side of the body.

35% felt depressed.

30% needed assistance to walk.

26% needed help with basic tasks like tying their shoes or eating

26% became residents at a nursing home

19% had difficulty speaking or understand others

Examples – Tim Curry, Gerard Ford, Hugh Hefner, Frankie Muniz, Woodrow Wilson, Richard Nixon, Kevin Sorbo, Charles Dickens, Franklin D. Roosevelt, and Dwight D. Eisenhower.

292. Stutter

AKA Stammer

A stutter is a speech disorder where the flow of speech is interrupted by involuntarily repetitions. Some stutterers elongate sounds, syllables, or words but don't repeat them. Some stutterers don't just repeat sounds but entire sentences. The specific sound that a stutter has difficulty saying is known as a Block.

Stuttering usually begins in childhood and can affect a person for two to five years. In 20% of cases, the stuttering continues into adulthood. Sometimes, it lasts a lifetime.

If you see a person stuttering, it is vital that you don't interrupt them or try to complete their sentence. Although a stutterer can take a long time to say one sentence, it's pivotal that the people around him or her are patient and allow them to speak without helping or rushing them.

Many people believe that stuttering is due to nervousness. In reality, stuttering is a neurological disorder. It can be due to genetics or an abnormality in the brain.

There are many ways to treat stuttering including medication, breathing techniques, and therapy.

Example – King George VI's stutter was depicted in the film, The King's Speech. Although the king's stutter seems to be treated by the end of the film, George VI struggled with his stammer for the rest of his life. Although treatment made him stutter less, he was never able to speak clearly for the rest of his reign.

293. Subjective Double Syndrome
AKA Doppelgänger Syndrome
Subjective Double Syndrome is the belief that there is a duplicate of you or someone else in the world. A person with this condition may believe the doppelganger is a long-lost twin, a clone, a phantom, a shapeshifter, or a being from a parallel universe.

There is a more extreme version of it called Clone Pluralization, where the person believes there are multiple copies of himself.

294. Sudden   Unexpected   Nocturnal   Death
        Syndrome
This syndrome causes a seemingly healthy person to suddenly die in their sleep with no visible symptoms leading up to their demise. It was first reported in the 1970s in Hmong males in Laos. SUNDS tends to affect men who had unknown heart conditions.

SUNDS occurred so often in Laos, it popularized the myth of Dab Tsuam (pronounced "da cho,") a nocturnal pressing spirit that sits on the victim's chest to stop them breathing. Dab Tsaum translates as "crushed by a ghost." Hmong men were so scared of Dab Tsuam that some of them went to sleep dressed as women, believing the spirit only preys on males.

SUNDS has also regularly occurred in the Philippines and Japan. The Philippines people call it Bangungot which means "to rise and to moan in sleep." The Japanese call it Pokurri which means "sudden and unexpectedly ceased phenomena."

No one knows what causes this illness.

295. Sudoku Attack

There are some disorders that are so specific, they have only happened to one person in history. An anonymous man survived an avalanche even though he was deprived of oxygen for 15 minutes. Although he survived, he suffered one major side effect – he has seizures but only if he does a Sudoku. After he learned the sudokus were his trigger, he quit doing the puzzles and hasn't had a fit since.

296. The Suicide Disease

AKA Trigeminal Neuralgia, Prosoplasia, Trifacial Neuralgia, Fothergill's Disease

A person with this disorder suffers unbearable pain to their fifth cranial nerve when they perform a specific trigger e.g. smile, kiss, shave, laugh, cry, etc. Since the nerve is near the mouth, many sufferers have their teeth removed unnecessarily, believing it will stop the pain. The pain feels like being stabbed and electrocuted in the face at the same time. The pain is so excruciating, 50% of sufferers end up taking their own life. Sadly, there is no cure.

297. Sun Allergy
AKA Xeroderma Pigmentosum
This genetic disorder prevents DNA to repair itself when exposed to ultraviolet light. Within minutes of being exposed to sunlight, the person will suffer severe sunburn, cracked skin, damaged nerves, hearing loss, and seizures. People with XP are prone to skin and brain cancer.

298. Super Beauty Syndrome
A mutation in the FOXC2 gene will cause a person to have a double set of eyelashes. This condition causes a lot discomfort since the eyelashes tend to scrape the eyes, which can lead to blindness. It also causes the person to have incredibly striking eyes.
Example - Legendary actress, Elizabeth Taylor, had this disorder. In fact, this condition helped her achieve super-stardom. The mutated gene made Taylor look so stunning, the makeup team for her films kept ignoring her because they assumed eyeliner and mascara had already been applied to her since her eyes were so striking.

299. Super Hearing Disorder
AKA Superior Canal Dehiscene, Autophony
In DC Comics, Superman's hearing is so acute, he can hear cortisol forming in a person's body. He can even hear cells dividing. Doesn't sound awesome? Not necessarily.

Justine Mitchell was 39 years old when she developed Super Hearing, which made her life "a misery." Her hearing was so acute, she could hear her own eyeballs moving in her skull. When Mitchell looked around, her eyes "sounded like sandpaper on wood in my head." She could hear her blood flowing through her veins, her intestines churning in her belly, and her heartbeat pounding like a drum. A coffee machine was so deafening, Mitchell could barely stand straight. She couldn't talk without feeling ill. Thankfully, Mitchell had an operation that cured her of her super-hearing.

300. Supernumerary Nipple
AKA Pseudomamma, Polythelia, Triple Nipple, Nubin
A supernumerary nipple is basically an extra nipple. Although most people with this condition only have one extra nipple, there have been reports of people having up to seven nipples in total. Some of the extra nipples can lactate. In rare occurrences, the extra nipple can develop on the face or foot. There has been one case of a woman forming an entire breast on her back.
Example – Harry Styles

301. Swine Flu
Like influenza, swine flu causes a person to develop a fever, a cough, decreased appetite, fatigue, and a coughing fit. If a pregnant woman catches swine flu, she can suffer a miscarriage. Although there was a panic when swine flu became public knowledge in 2009, it has a mortality rate of 4% at most.

Swine flu is descended from the Spanish flu that swept the world in 1918.

The Center for Disease Control originally referred to swine flu as novel influenza since they were afraid that the knowledge the disease originated from pigs would devastate the pork industry. Although infected pigs can spread the virus to birds, humans, and other pigs, most humans caught the disease from other people.

Developing swing flu causes the person to have an incredible resistance to regular flu.

Although there is medication for swine flu, most people infected with the illness make a recovery without receiving any medical attention.

302. Syndrome of Delusional Companions

A sufferer of SDC believes that objects have self-awareness and emotion. Strangely, this was a common belief in Ancient Greece known as animism.

Example – In the 2007 film, Lars and the Real Girl, the title character is convinced his sex doll is alive.

303. Synesthesia

This disorder is when stimulation for one sense leads to the involuntary stimulation of another sense. A synesthete might taste chocolate every time they hear music from a violin. A synesthete may smell mustard when they touch rubber. Synesthesia is usually permanent. Most synesthetes are unaware that what they are suffering is a disorder since they are oblivious that other people aren't experiencing the same symptoms. There are many versions of this disorder.

i)   Number Form causes the person see numbers

every time they think of them.

ii) Grapheme-Color Synesthesia makes letters and numbers tinged with color.

iii) Auditory-Tactile Synesthesia causes a certain sound to trigger a different sensation e.g. Hearing a car siren can cause a synesthete to feel pain in the left arm or intense cold in their legs.

iv) Misophonia makes a synesthete become hysterical or enraged if they hear a specific sound e.g. the sound of a text or the tapping of a keyboard. The 2016 documentary, Quiet Please, gives a detailed look on the condition.

v) Lexical-Gustatory Synesthesia causes the affected person to experience tastes when hearing certain sounds.

vi) Mirror-Touch Synesthetes experience the same sensation as another person. If you had this disorder, you might feel like you have been punched in the face after watching someone else being hit. People with this disorder tend to be far more empathetic than normal.

vii) Chromesthesia causes a person to see colors or bright lights when exposed to certain sounds. A synesthete called Tori Amos could play music on the piano after she heard it once when she was two years old. Amos could compose her own music on the piano when she was three. Chromesthesics often believe they are having visions or seeing people's "auras," since they are oblivious they are suffering a type of synesthesia.

304. Syphilis
AKA Great Pox
Syphilis is a disease that occurs in four stages –

i)   Primary syphilis causes a painless, non-itchy rash and multiple sores.
ii)   Secondary syphilis causes the rash to spread to the palms of the hands and the soles of the feet.
iii) Latent syphilis has little to no symptoms for several years
iv)   Tertiary syphilis develops non-cancerous growths called gummas.

Although syphilis is known as a sexually transmitted disease, you can catch it by touching one of the sores on a syphilitic person's skin.

Since the disease is associated with sex, syphilis was often used to slander others. The Russians called it "the Polish disease." The Polish called it "the German disease." The Portuguese called it "the Spanish disease."

Syphilis is nicknamed The Great Imitator as its symptoms causes it to be mistaken for many other diseases.

To date, there is only one known cure – malaria. If an infected person develops malaria, their temperature will skyrocket, killing the disease. However, the survival of the patient is not guaranteed. Even if the person does survive, he or she may be left sterile. Although syphilis is nowhere near as devastating as it used to be, it still kills about 100,000 people per year.

305. Tanganyika Laughter
In 1962, over a thousand people suffered fits of

laughter in Tanganyika (now known as Tanzania.)

But they didn't laugh for a few hours or days. Some of them laughed hysterically for months. Several of them had violent outbursts for up to two weeks. The epidemic got so bad, 14 schools were forced to closed.

306. Teratomas

Teratomas is when a tumor develops another organ. Teratomas can develop bones, hair, and eyes. "Teratomas" is Greek for "monster swelling."

In 2014, four fully-formed teeth were found in the brain of a four-month old baby in Maryland, US. This is the only time this has been recorded in human history. After the teeth were removed, the infant made a complete recovery... somehow.

Although this might sound like something from a horror movie, it has a simple explanation. A tumor is a collection of cells that are dividing at an uncontrollable rate. Some bone cells might separate from where they are meant to be and continue developing bones in other body parts.

307. Textiety

This is a clinical condition where a person suffers anxiety because someone won't text them back.

308. Thought Insertion

This is the belief that thoughts are being inserted into a person's mind against their will. The sufferer often believes these thoughts are inserted by the government through microchips, radio waves, satellites, or through a television screen.

Example – Serial killer, Aileen Wuornos, was adamant that the government was putting thoughts in her mind by using specific frequencies.

309. Tinnitus

Tinnitus is the hearing of a sound when there is no external noise to be heard. The sound can be a ring, a click, a hiss, or a roar. The tinnitus sufferer may hear the sound from one ear or both. The sound can be low or high in volume and pitch. Tinnitus is usually a symptom of hearing loss, a brain tumor, or a head injury.

Suffering tinnitus for years can cause a person to develop anxiety or depression as they literally can't have a second of peace.

If you are in a situation where you regularly hear a loud noise e.g. traffic, loud music, another person snoring, etc. it's imperative that you use ear plugs or some sort of noise-cancelling device to avoid developing tinnitus.

Sadly, there is no cure for tinnitus and treatment against it is ineffective.

310. Tonsillitis

Tonsillitis is an inflammation of the tonsils, which causes the lymph nodes in the neck to swell. When the tonsils are enflamed, it can be extremely painful to eat, drink, or swallow one's saliva. 7.5% of people develop tonsillitis. To avoid developing an abscess, the tonsils need to be surgically removed. Astoundingly, this procedure has been carried out for 3,000 years.

During surgery, 80% of the tonsils are removed. In rare cases, the tonsils can grow back.

Because tonsils are one of the body's first responders to pathogens, it is more likely a person will develop viruses after their tonsils are removed.

Garlic, lemon, salt water, and grapefruit are recommended when one is suffering tonsillitis since they are easy to eat while the throat is sensitive.

311. Tooth Decay
AKA Dental Caries, Cavities
Tooth decay is caused by excessive sugar, refined carbohydrates, and acid-producing bacteria.

Tooth decay is the most common illness in the world apart from the common cold. Although it affects 25% of children from 2-5, it affects 90% of adults over the age of 40. Since the developed world has effortless access to sugary food, tooth decay is more common in First World Countries.

Many people don't know that tooth decay is contagious. If you kiss someone with cavities, the bacteria can spread to your mouth and develop into gum disease. It's also possible to spread cavities by sharing food or kitchen utensils.

Although many people dismiss toothache because they know that going to the dentist is usually a painful experience, it's vital since tooth decay can be fatal. An infection in an upper back tooth can spread to the sinus, which can lead to the brain resulting in death.

Tooth decay can be prevented by having a low-sugar diet, flossing, and brushing your teeth.

312. Tourette's
This neuropsychiatric disorder causes a person to develop several motor tics and at least one vocal tic.

Tics are sudden twitches, movements, or sounds. A physical tic manifests as an involuntary jerky movement in the shoulders, arms, neck, or face.

A vocal tic can sound like a sniff, laugh, scream, gurgle, sound effect, or a word. Although many people believe that Tourette sufferers regularly swear or say inappropriate things, this isn't true. The impulse to say inappropriate things is a separate neurological disorder called coprolalia. 10% of Tourette's sufferers have this disorder. Sadly, there is a 79% chance that a Tourette's sufferer will have another neurological disorder such as ADHD, OCD, anxiety, or dysgraphia.

Tourette's is usually noticeable during childhood. However, the tics that a child has can change in severity or complexity as the they get older.

Males are three or four times more likely to develop Tourette's than females.

It's straining and unhelpful for a Tourette's sufferer to suppress a tic.

Despite what many people believe, Tourette's has no effect on a person's intelligence or life expectancy.

Although the tics can subside from a time, Tourette's is a lifelong disorder. Sadly, there is no cure for the condition.

313. Tree Man Syndrome
AKA Epidermodysplasia Verruciformis
A man called Dede Koswara developed EV after cutting his knee. This developed into an infection called human papillomaviruses (HPV.) Because Koswara had a weak immune system, the virus consumed his skin cells, causing them to develop into gnarled branch-shaped growths. Although he had 95%

of them removed, they grew back worse than ever within a few months. Sadly, there is no known cure.

314. The Truman Show Delusion

A person with this condition believes they are being watched by people and secret cameras. Obviously, the condition's name is based on the film, The Truman Show, which revolves around a man who learns his entire life is a tv show and he lives in an enormous soundstage. Everything is scripted including who is he going to marry and who his friends are. To make sure Truman never tries to leave the country, the director makes Truman's "father" drown, making Truman have a fear of travelling by sea.

Dr. Joel Gold treated a man with Truman Show Delusion who believed that the 9/11 attacks were faked to scare him into not flying anywhere to escape the city-sized soundstage.

315. Tuberculosis

AKA White Plague, King's Curse, Scrofula, Phthisis

TB is a lung infection that causes a person to develop a chronic cough, fever, and weight loss. The most well-known symptom is coughing up blood. TB used to be called consumption because the sufferer lost so much weight, it was like the disease was consuming them. The word is derived from "tuber" which is Latin for "lump" of "swelling."

TB is air-born and highly contagious. Although TB begins in the lungs, it can spread through the central nervous system, the lymph nodes, and the bones.

Nowadays, TB is diagnosed through chest x-rays or by scanning bodily fluids under a microscope.

TB is one of mankind's oldest diseases. Although it has been around for over 40,000 years, it began to infect humans about 5,500 years ago. Although TB was a terrifying illness, a vaccine was produced in 1921. Despite the fact that this has caused TB infections to plummet worldwide, it still kills over a million people per year, making it the most lethal infectious disease in the world.

Examples – President James Monroe, Emily Bronte, Anton Chekhov, Henry VII, Alexander Pope, Eleanor Roosevelt, John Keats, Louis Braille, Franz Kafka, Vivien Leigh, and George Orwell.

The disease was depicted in the films, Moulin Rouge and Les Misérables.

316. Typhoid Fever

When the salmonella typhi bacterium infects a person, he or she will develop symptoms within 6-30 days. These symptoms include fever, constipation, headaches, and abdominal pains. Some people develop a rash covered with rose-colored spots.

Although it's hard to decipher typhoid in its early stages, a medical journal published in 1976 stated that typhoid sufferers have a smell that resembles baked bread.

The word "typhoid" means "resembling typhus" since both disorders have similar symptoms.

In 2015, 12.5 million people were recorded with typhoid, mostly in India. Without treatment, a typhoid sufferer has a 20% chance of dying.

Nowadays, typhoid is treatable with basic medication. The most extreme cases can be rectified with drainage surgery.

Example – Typhoid become well-known after a cook known as Typhoid Mary (whose real name was Mary Mallon) infected 51 people with the disease, killing three. When she was quarantined on March 27th 1915 by New York health officials, the doctors learned something astounding. Although Mary carried the typhoid bacterium, she was immune to it. A person who is immune to the pathogen that they carry is known as an asymptomatic carrier. This was depicted in the 2014 television series, The Knick.

317. Ulcers
An ulcer is a break in a bodily membrane, which impairs the neighboring organs to function.

Until 1984, it was believed that ulcers were caused by stress. A gastroenterologist called Barry Marshall proved that ulcers are caused by bacteria. Antibiotics are the best treatment for ulcers.

318. Unbreakable Bone Syndrome
The film, Unbreakable, revolves around a man called David Dunn who is the sole survivor of a train crash. Dunn realizes that he has never had an injury in his life and the only time he has nearly died is when he drowned. Bizarrely, there is a man, simply called John who has been in a very similar situation. John has a mutation so his bones are eight times stronger than a normal person. When John was involved in a car crash in 1994, he didn't break a single bone. This disorder seems to be genetic as his children have never broken a bone despite being involved in accidents that should have caused serious injuries.

Although this sounds fantastic, John has one huge

weakness – water. His bones are so heavy, it's impossible for him to swim. He also finds it exhausting to walk as he has to drag around his incredibly heavy skeleton.

319. Unlimited Color Syndrome
AKA Tetrachromat Syndrome
The three types of cones in a person's eyes allows them to see seven million colors. However, some women have an extra cone, which allows them to see a hundred million colors. A person with an extra cone is called a tetrachromat. This means that tetrachromats can see colors that other people can't imagine.

320. Unstoppable Syndrome
AKA Congenital Analgesia
This condition causes the person's muscles to harden to a point where they are immune to pain. Although this sounds like a superpower, it often leads to the sufferer dying prematurely. If they touch something hot or sharp, their body doesn't send a signal to alert them of the damage. This often leads to horrific infections that develop into gangrene or necrosis.
Example – The villain, Niederman, in the film, The Girl Who Played with Fire, has this condition.

321. Vampire Syndrome
VS causes the sufferer to resemble a vampire. The affected person has no sweat glands and has thin skin that lacks pigmentation. Anyone with VS must avoid sunlight or they will overheat within a matter of minutes. VS causes certain teeth to fall out, giving the impression that the affected person has fangs. It has a

similar effect on the hair, giving the person a deranged hairstyle. There is no known cure.

322. Vertigo

Vertigo is a feeling of spinning or swaying. Despite what is depicted in the film, Vertigo, the condition is not a fear of heights. You can suffer vertigo while being on the ground floor of a house. Vertigo can leave a person so disoriented that they will sweat, vomit, or lose the ability to walk.

Vertigo can develop from a stroke, trauma, brain tumor, carbon monoxide, alcoholism, or a bad reaction to aspirin.

323. Vitiligo

Vitiligo is a skin condition that causes the pigment cells in the skin, melanocytes, to be destroyed. These cells usually die around the mouth, eyes, and nose. Vitiligo also causes hair to go prematurely grey. Although it effects all races, people associate vitiligo with Africans since the condition stands out more on darker skin.

Vitiligo has no effect on the texture of the skin nor does it cause any pain. However, it does make the sufferer more susceptible to sunburn and skin cancer.

Doctors are still unsure what causes vitiligo but genetics is a factor since there's a 20% chance that a person suffering vitiligo will pass it on.

Example – Michael Jackson, Jon Hamm, Steve Martin

324. Walking Corpse Syndrome
AKA Cotard Delusion

This causes the affected person to believe they are

dead or their organs are rotting away. They might see themselves as a ghost or a zombie. According to Jules Cotard, who originally hypothesized the disorder, 55% of patients paradoxically believe they are immortal.

Some people with this condition think that they never existed and are simply a figment of another person's imagination.

Example – In Season 4 Episode 14 of the tv series, Scrubs, the doctors interact with a person with this syndrome.

325. Wart

A wart is a rough, hard growth on the skin. Despite what urban legends may have you believe, warts are not caused by toads. A wart is caused by the human papilliomavirus (HPV.) Although some sources say warts are contagious, this is not true.

A veruca is a plantar wart that occurs on the foot. They become painful when pressed which makes walking difficult.

Although you can freeze a wart to make it go away, they usually disappear by themselves after a few weeks. However, the most effective thing against warts is… duct tape. I swear that's true.

326. Water Allergy
AKA Aquagenic Urticaria

This condition makes a person allergic to water. Now, you might think – Don't you need water to live?

Yes. Yes you do. But this is worse than it sounds. Not only can you not drink water, you can't touch it. If you sweat or a single tear leaks out of your eye, your body will be covered in an agonizing rash. Only 40 people in the world have this incredibly rare disorder.

Oral antihistamine, phototherapy, barrier cream, and stanozolol are recommended if you happen to be one of the 0.0000002% of people on Earth who suffer this.

327.   Water on the Brain
AKA Hydrocephalus

This condition forces cerebrospinal fluid to build up in the brain, which causes pressure in the skull. This results in impaired balance, double vision, mood swings, migraines, incontinence, and mental impairment. If it occurs in babies, the head will noticeably increase in size and the infant can be prone to vomiting and seizures. WOTB can be caused by meningitis, a brain tumour, a head injury, or a haemorrhage.

The first device to work successfully to treat WOTB was called the Wade-Dahl-Till valve. It was created by Stanley Wade, Kenneth Till, and... the author of Charlie and the Chocolate Factory, Road Dahl! It was the first device of its kind that didn't jam, minimising the possibility of brain damage.

Nowadays, the disorder is treated by inserting a small shunt in the person's skull, allowing the liquid to be drained from the brain.

328.   Wendigo Psychosis

Wendigo Psychosis causes the sufferer to develop a craving for human meat. Although this disorder is supposed to have originated with Native Americans, no one has seen an irrefutable case of WP. In fact, many historians believe the Wendigo Psychosis isn't a real disorder.

329. Werewolf Syndrome
AKA Hirsutism
This causes a person to develop excess body-hair in places where hair shouldn't normally grow e.g. forehead. A man with this disorder usually have his entire face covered in hair so he looks like a wolfman. Many historians believe the werewolf mythos was created because people misunderstood that hirsutism was a condition, not a curse.

330. Williams Syndrome
A person with Williams Syndrome suffers from poor visual-spatial ability and heart defects. Weirdly, they always come across as being astoundingly happy.

331. Xhosa Nerves
AKA Amafufunyana
This is a culture-bound syndrome which means it only effects a specific region. XN mainly effects the Xhosa people but has been documented to affect members of the Zulu tribe.

When a person has schizophrenia, they might hear a voice that isn't there.

However, a XN sufferer hears a voice... from their stomach. Although this sounds silly, Xhosa find this disorder terrifying. Xhosans who have suffered this disorder had their stomach tell them to commit suicide, jump in front of a vehicle, or threaten them with seizures.

The stomach can talk to the sufferer all day,

making them prone to nightmares and fatigue. Although Xhosans believe XN stems from a curse, doctors aren't certain what causes the illness.